THE ANTHRAX LETTERS

A Medical Detective Story

THE ANTHRAX LETTERS

A Medical Detective Story

Leonard A. Cole

Joseph Henry Press
Washington, D.C.

Joseph Henry Press • 500 Fifth Street, NW • Washington, DC 20001

The Joseph Henry Press, an imprint of the National Academies Press, was created with the goal of making books on science, technology, and health more widely available to professionals and the public. Joseph Henry was one of the founders of the National Academy of Sciences and a leader in early American science.

Library of Congress Cataloging-in-Publication Data

Cole, Leonard A., 1933-
The anthrax letters : a medical detective story / Leonard A. Cole.
p. cm.
Includes bibliographical references and index.
ISBN 0-309-08881-X — ISBN 0-309-52584-5 (PDF)
1. Bioterrorism—United States. 2. Anthrax—United States. 3. Postal service—United States. 4. Victims of terrorism—United States. I. Title.
HV6432.C63 2003
364.152′3—dc22
2003015149

Printed in the United States of America.

CONTENTS

PROLOGUE

When Pat Hallengren arrived at work on August 10, 2002, she noticed that the middle mailbox was missing. It was the one she had always used. For as long as she could remember, it had stood between two other receptacles outside her window in the American Express travel office in Princeton, New Jersey.

But during the late hours of the previous night, postal authorities had removed the box. Before long, word spread about the reason, and local curiosity turned to horror. The mailbox was found to have contained anthrax spores. When Pat heard this, her first thoughts were about her mailman, Mario. "I really wasn't concerned for myself. I mean, I just put mail in the box, but Mario had to take it out." Her worry was understandable.

Anthrax bacteria are as murderous as South American flesh-eating ants. An army of ants, traveling in the millions, can decimate an immobilized individual by devouring his flesh layer by layer. Death is gradual and agonizing. Anthrax bacilli do to the body from within what the ants do from without. They attack everywhere, shutting down and destroying the body's functions from top to bottom. The organisms continue to multiply and swarm until there is nothing left for them to feed on. In 2 or 3 days a few thousand bacilli may become trillions. At the time of death, as much as 30 percent of a person's blood weight may be live bacilli. A microscopic cross section of a blood vessel looks as though it is teeming with worms.

The anthrax bioterrorism attacks the previous fall, in 2001, had been conducted by mail. On October 4, three weeks after the terror of September 11, a Florida man was diagnosed with inhalation anthrax. His death the next day became the first known fatality ever caused by bioterrorism in the United States. During the following weeks, more people were diagnosed with inhalation anthrax as well as with the less dangerous cutaneous, or skin, form of the disease.

Almost all the cases were traced to spores of *Bacillus anthracis* that had been placed in letters. Perhaps a half dozen letters containing a quantity of powder equivalent in volume to a handful of aspirin tablets paralyzed much of America. During the fall 2001 scare, congressional sessions were suspended and the U.S. Supreme Court was evacuated. Infected mail disrupted television studios and newspaper offices. People everywhere were afraid to open mail.

Four of the anthrax letters were later found, and all were postmarked "Trenton, NJ." That was the imprint made at the large postal sorting and distribution center on Route 130 in Hamilton Township, 10 miles from Princeton. Ten months after the attacks, when Pat Hallengren's favorite mailbox had been removed, mailboxes that served the Hamilton facility were belatedly being tested for anthrax. In the first week of August, investigators swabbed 561 drop boxes and delivered the cotton tips to state laboratories. Only that one mailbox, on Nassau Street near the corner of Bank Street in Princeton, tested positive for anthrax. Could that box, not 30 feet from Pat Hallengren's desk, have been where the poison letters were deposited?

The mailer of the anthrax letters had not yet been found. But 6 weeks before the discovery of anthrax spores in the Princeton mailbox, the Federal Bureau of Investigation had identified a microbiologist named Steven Hatfill as a "person of interest."

Days after the middle mailbox was removed, federal agents fanned out through the neighborhood. They showed a picture of a steely, thick-necked man to merchants and patrons up and down Nassau Street. It was Hatfill. "Do you remember seeing this person?" they asked. "I don't recognize him," Pat Hallengren answered, "but I see so many people on this corner." Four doors up from the corner, Shalom Levin, the bearded owner of the Red Onion delicatessen, was ambivalent. "I might have seen him walking around here," he told an FBI official. But perhaps Hatfill's face seemed familiar, he acknowledged, because he had seen it on TV. In 2003, long after the discovery of anthrax in the Princeton mailbox, the FBI was still searching for the mailer and still considered Hatfill a person of interest.

Between October 4 and November 21, 2001, 22 people were diagnosed with anthrax. Eleven contracted the cutaneous form and all survived. But among the 11 who became ill from inhaling spores, five died. In subsequent months, with no new cases, national anxiety eased. But the discovery of the contaminated mailbox almost a year later in Princeton drew a torrent of television and newspaper coverage from around the world. Fear had been rekindled.

Concern about anthrax is as old as the Bible. Primarily a disease of animals, it is thought to have been the fifth of the 10 biblical plagues visited by God on the ancient Egyptians for refusing freedom to the Jews. As recounted in Exodus, horses, donkeys, camels, cattle, and sheep were struck "with a very severe pestilence." After their carcasses were burned, the virulence of the anthrax germs persisted, for the soot caused "boils on man and beast throughout the land of Egypt."

In recent years, anthrax spores have been deemed among the most likely of biological weapons because they are hardy, long lived, and, if inhaled, utterly destructive. A victim is unlikely to know he is under attack. As with other biological agents, anthrax germs are odorless and tasteless, and lethal quantities can be so tiny as to go unseen.

Every 3 seconds or so, a human being inhales and exhales about a pint of air. Each cycle draws in oxygen to fuel the body and releases carbon dioxide, the gaseous waste product. The inhaled air commonly carries with it floating incidentals such as dust, bacteria, and other microscopic particles. If a particle is larger than 5 microns, it is likely to be blocked from reaching deep into the lungs by the respiratory tract's mucus and filtration hairs. If smaller than 1 micron, a particle is too small to be retained and is blown out during exhalation. An anthrax spore may be 1 micron wide and 2 or 3 microns long, just the right size to reach deep into the respiratory pathway.

A spore is so tiny that a cluster of thousands, which would be enough to kill someone, is scarcely visible to the naked eye. A thousand spores side by side would barely reach across the thin edge of a dime. Once inhaled, the spores are drawn into the bronchial tree where they travel through numerous branches deep in the lungs. Near the tips of the branches are microscopic sacs called alveoli. It is in these sacs that inhaled oxygen is exchanged with carbon dioxide.

Stationed among the alveoli are armies of defender cells called macrophages. These cells sense foreign microinvaders and engulf them. A pulmonary macrophage normally destroys its inhaled captive and taxis it to the lymph nodes in the mediastinum, the area between the lungs. But in the case of anthrax, spores may transform into active, germinating organisms before the macrophage can affect them. The bacteria then can reproduce and release toxin that destroys the macrophage. Thus, in a perverse turnabout, the anthrax bacteria, like soldiers in the Trojan horse, can burst out of their encirclement, into the lymph and blood systems.

An infected person at first is unaware that a gruesome cascade is under way. Although the onslaught is relentless, symptoms do not appear immediately. Fluids that have begun to accumulate in the mediastinum gradually pry the lungs apart. Breathing becomes increasingly difficult, and after a few days a person feels as if his head is being held underwater, permitted to bob up for a quick gulp of air and then pushed under again.

The agony works its way through the body. Nausea gives way to violent, bloody vomiting. Joints are so inflamed that flexing an arm or leg becomes an act of torment. Bloody fluids squeeze between the brain and skull, and the victim's face may balloon out beyond recognition. The tightening vice around the brain causes excruciating pain and delirium. Survival depends on being provided appropriate antibiotics before the bacteria have released so much toxin that the body cannot recover. If inhalation anthrax is not treated in time, almost all victims suffer a tortured death. One organ after another is decimated—the lungs, the kidneys, the heart—until life is sucked away.

It is because of such ghastly effects that anthrax and other biological agents have been prohibited as weapons by international agreement. The treaty that bans their development or possession by nations, the 1972 Biological Weapons Convention, uniquely describes their use as "repugnant to the conscience of mankind." Yet despite this widely accepted moral precept, a germ weapon is seen by some not as a shameful blight but as a preferred instrument of terror.

THE ANTHRAX LETTERS

CHAPTER
ONE

DEADLY DIAGNOSIS

Bob Stevens wore a huge smile. He had just reached the observation deck with his wife, Maureen, and daughter, Casey, after a 26-story elevator ride through Chimney Rock, a mountain of granite. Before them lay the stunning expanse of North Carolina's Chimney Rock Park. Far below and to the right was Lake Lure, long and calm. To the left stood Hickory Nut Falls, where icy water cascaded 400 feet down into a valley of rocks. Seventy-five miles away toward the horizon loomed King's Mountain. "I feel at peace with the world," Bob said.

September 28, 2001, a Friday, was a perfect day to visit the park. The horror of September 11 was still raw, and an afternoon of natural splendor would be a pleasant distraction. The previous day Bob and Maureen had driven 11 hours from their Lantana, Florida, home to visit Casey in Charlotte. Then on Friday, after breakfast in Casey's apartment, they drove west and picked up Route 74 toward Asheville. Ninety minutes later they were in the park. The sky was clear and the autumn air fresh. Color was everywhere—purple mountain flowers, aqua green lichen, red and orange oak leaves. As they walked along a trail, Bob detoured past some rocks to a waterfall. He cupped his hands, reached in, and drank two good scoops of water. Later, on the way back to Charlotte, they had dinner at an Italian restaurant before settling into Casey's place for the evening. They were still giddy about the mountains and the changing colors.

For Robert Stevens, 63, a veteran photo editor, the scenes in the park were especially enthralling. They fed his esthetic appetite in a way that the two-dimensional images he worked with could not. Still, Stevens enjoyed his job at the *Sun*, a supermarket tabloid published in Boca Raton, 20 minutes from Lantana. Like its half dozen sister publications owned by American Media, Inc., the *Sun* specialized in sensationalism. Bob had worked for one or another AMI tabloid, including the more famous *National Enquirer*, since emigrating from England in 1974. He had tried retirement in 2000 but missed his job and fellow workers. Back at his desk the next year, he delighted in servicing readers who liked stories about psychics and seers and pig races. So what if some tales were bizarre or exaggerated? The pictures he retouched had their own odd esthetic appeal—female Elvis impersonators, women who lost weight through prayer.

Bob's puckish humor seemed suited to the amusing themes he worked on. But he was serious about his craft. "The best in the business," judged Lee Harrison, a fellow expatriate from England who had worked with Bob at the *National Enquirer*. "He's brilliant on the computer, great at touching up photographs to make celebrities look good. Unless of course it was a story about a celebrity not looking too good," Lee chuckled.

Bob and Maureen were married in 1974 before leaving for the United States. Casey, the youngest of their four children—the other three were from previous marriages—was "the apple of his eye," Maureen would say. At 21, Casey had recently found a position as an actuary in Charlotte, and her parents were thrilled to see how well she had settled in. Laughs, jokes, and hugs, long a mainstay of Bob's life, were abundant between father and daughter. The day after they visited the park, they strolled around downtown Charlotte. The weather was windy and chilly, and they spent much of the time in an indoor mall before lunching at an Irish pub. Bob had a penchant for pubs. He frequently stopped for lunch at the Lion and Eagle in Boca Raton or the Blue Anchor in Delray Beach. Both had a clientele of English expatriates like himself. Smoked salmon on brown bread, a pint of Harp, and trading jokes with friends were a favorite way to break the day.

After seeing the office building where Casey worked, they drove back to her apartment. It was late afternoon, and Bob felt uncharacteristically tired. A biking and hiking enthusiast, he was usually brimming with energy. But now he felt he had to rest while his wife and daughter went shopping. He wanted to be in good shape for the next day's trip to Durham, where he would be meeting Casey's boyfriend, a student at Duke University. Maureen and Casey returned to the apartment at 7:30 p.m. for a dinner of warmed-up leftovers. Bob joined them but, still fatigued, went to bed soon after.

The next morning, Sunday, Bob seemed better, and after Maureen and Casey returned from church they set out for Durham. But midway through the $2^1/_2$-hour drive, he began to shiver and shake. They stopped so he could climb into the back seat. "Your face is red," Maureen observed. Casey began to worry. "Dad, let's turn around and go home." "No, no," Bob insisted, "I'm not going to spoil your day."

Casey's boyfriend greeted them at his fraternity house and, when he saw how badly Bob was feeling, took him up to one of the bedrooms to lie down while the others went out for lunch. When they returned at 3:30, Bob was still weak and feverish, so he, Maureen, and Casey decided to go back to Charlotte right away. As they passed the university hospital, Casey and Maureen urged him to stop at the emergency room. He refused, and he continued to refuse as they passed other hospitals and clinics on the return trip. Bob fidgeted in the backseat, alternately sitting and lying down. Sighing frequently, he tilted his head back, saying it helped him breathe more easily. After arriving at Casey's, he nibbled at dinner, said he wanted to leave for home the next morning, and went to bed. Maureen and Casey stayed up talking and worrying before themselves going to sleep.

When Maureen awoke she saw that Bob's pillow was soaked with perspiration, but he said he felt all right. They packed the car, hugged Casey goodbye, and were on the road by 6 a.m. Bob insisted on taking the wheel and Maureen agreed to navigate. Sipping frequently from a bottle of water seemed to help Bob recover energy. They stopped once for a brief rest and then for gasoline near Jacksonville, about 300 miles from their destination. They reached Lantana about 5 p.m., and Bob pulled the white Saturn into the driveway of their ranch-style house. After unpacking and sorting the mail, Bob had a turkey sandwich with some hot tea.

Bob and Maureen both felt depleted and took their temperatures. Bob's was 101, Maureen's 102. "We've come down with a bug," Maureen thought. Bob went to bed at 8 p.m., and Maureen followed a bit later. Some time after 1 a.m. she awoke to the sound of retching. Bob had vomited in the bathroom and then come back to bed. Maureen noticed that he was fully dressed. She asked him how he felt, and he responded incoherently. She was annoyed with herself for not making him see a doctor sooner. Now she insisted.

Maureen threw on some clothes and helped Bob into the car. JFK Medical Center in Atlantis was only a mile away, and they arrived there at 2 a.m. On admission to the emergency department, Bob seemed delirious. In the words of his medical case description, "he was not oriented to person, place, or time." Clearly his brain was under some sort of stress. Maureen reviewed his behavior and symptoms for the past few days with the emergency room doctor. After an initial examination the presumptive diagnosis was meningitis, an inflammation of the membrane covering the brain. Although the causes of meningitis are various, one common source is bacterial infection. Accordingly, Bob was started on multiple antibiotics—cefotaxime and vancomycin—in addition to medication for nausea that made him sleepy. Maureen herself displayed none of Bob's symptoms. Her earlier fever proved to be transient and unrelated to Bob's.

Around 5:30 a.m. the emergency room staff prevailed upon Maureen to go home and rest for a few hours. When she returned to the hospital at 8 a.m., she learned that Bob had suffered a seizure and been intubated—a tube was threaded through his nose into his respiratory passage. The tube was attached to a ventilator, a device to help him breathe. A spinal tap had also been performed to examine his cerebrospinal fluid for signs of infection or other abnormalities.

Dr. Larry Bush, an infectious disease specialist, is an amiable skeptic with a fondness for conspiracy theories. Medical lore holds that when you hear hoofbeats, think horses, not zebras; that is, when trying to come up with a diagnosis, don't start with remote possibilities. Larry Bush is something of a zebra man. He is inclined toward less conventional thinking. He doubts, for example, the official line that President Kennedy was assassinated by a lone gunman shooting from the Dallas Book Depository building. "I stood at the site and some bullet trajectories line up with the grassy knoll." He smiles and shrugs as if to say, "That's what I think, so shouldn't I say it?"

Whatever one thinks of his Kennedy assessment, the mind-set of this slightly built physician with a gray-brown beard and mustache helped him point to the diagnosis of a lifetime. Bush began practicing out of the JFK Medical Center in 1989, a year after completing specialty training at the Medical College of Pennsylvania in Philadelphia. At 8:30 a.m. on Tuesday, October 2, 2001, he was in his office a few blocks from the hospital. He was about to leave for a meeting at the hospital when the phone rang. "We've got a 63-year-old man here with fever and apparent meningitis," a laboratory technician said. "His cerebrospinal fluid is cloudy and we did a Gram stain. We'd like you to look at it." Bush shot back, "I'm on my way. See you in a few minutes."

Cerebrospinal fluid, which runs through the brain and spinal column, is normally clear, like water. A cloudy sample, obtained through a spinal tap, suggests the presence of white blood cells, an indication of infection. The process of identifying the bacteria causing an infection commonly begins with a Gram stain. Introduced in 1844 by the Danish bacteriologist Hans Christian Gram, the test involves staining bacteria with crystal violet, a coloring agent that he developed, and then washing them with alcohol. Bacteria tend to fit into one of two categories according to whether they retain the violet color or not. Those that do—for example, bacteria in the genus *bacillus*, *clostridium*, *streptococcus*, or *staphylococcus*—are deemed Gram positive; those that do not are Gram negative.

When Bush arrived at JFK Medical Center he went directly to the laboratory. He looked into the microscope and then at the patient's record. His undergraduate degree was in microbiology, "so I tend to think like a microbiologist as well as a physician," he says. The Gram stain was positive, and the shape of the bugs amid the white cells suggested they were bacilli of some sort.

Dr. Bush went next door to the emergency department, where Bob Stevens lay unconscious. He introduced himself to Maureen, quizzed her briefly about what Bob had been doing the past few days, and examined him. He saw no skin lesions or indications of trauma and felt no swollen glands. Through the stethoscope he heard a crackling sound when Bob breathed, caused perhaps by an obstruction. It was clear that Bob was very ill, but why? Bush returned to the lab, looked again at the microscopic rods, and played out some thoughts:

> The organism was a bacillus, as evident by its Gram stain and its shape. There are many types of bacilli, but very few cause significant disease. Also, bacilli can appear in some blood samples as contaminants. But if you see bacteria in a normally sterile area like spinal fluid, you have to think of it as an infection, not a contaminant.

> So what I saw was obviously a bacillus in the spinal fluid. When you think of the common bacilli that can cause somebody to be ill, there is one called *Bacillus cereus,* which you can see with traumatized patients or with immuno-compromised patients. Another is *Bacillus subtilis*, which, again, we occasionally see in the bloodstream. I've never seen it in the spinal fluid. So my thought was that, although this could be a couple of these or some other bacilli, usually people who have them have a reason to have them. This patient had no reason to have any bacillus as far as exposures or trauma were concerned. He had not been an ill person and he had no immune system defects.

Larry Bush, like other physicians, indeed like much of America, had been hearing a lot about biological weapons in recent years. National concern heightened in the 1990s with suspicions that Iraq still had a biological warfare program despite its agreement after the Persian Gulf War to end it. Equally shocking was news about the size of the former Soviet program. In violation of the 1972 Biological Weapons Convention, a treaty that bans these weapons, around 60,000 Soviet scientists and technicians had produced tons of anthrax, smallpox, and plague germs. Although the program ended with the dissolution of the Soviet Union in 1991, Soviet scientists were subsequently courted by Iran and perhaps Syria, Libya, and other countries deemed unsavory by the American government.

Besides bioagents as military weapons of war, biological terrorism has become more worrisome. The 1995 release of sarin in the Tokyo subway by the Aum Shinrikyo cult was particularly alarming. In that case the weapon was a *chemical* nerve agent that killed 12 people and injured more than a thousand. But the attack suggested that the same, or worse, could be done with a *biological* agent. So by the beginning of the 21st century, bioweapons were understood to be a growing threat. Then came September 11. After the jetliners crashed into the World Trade Center and the Pentagon, terrorism became the subject overshadowing all other issues. Residents of South Florida were acutely aware that several of the suicide hijackers had lived among them and taken flying lessons nearby. Anxiety there as elsewhere was about further hijackings and other forms of terror, though the possibility of bioterrorism received no special attention.

For Larry Bush on October 2, the bioconnection was immediate. He turned to the woman who had prepared the Stevens specimens: "Kandy, did you see his cerebrospinal fluid? Did you look at that Gram stain?" Kandy Thompson, a medical technologist since 1975, had worked at JFK for 6 years. She had previously run a microbiology laboratory at another hospital, and Bush knew she could offer an experienced impression. She looked again through the microscope and said: "The spinal fluid looks milky, pussy, bloody. I don't know—large rods, a bacillus, maybe clostridium." Neither she, the emergency room doctors, nor anyone else who had attended Robert Stevens imagined the kind of thought that had leaped to Larry Bush's mind: "They're large. Really. You know, it could be anthrax," he said. Her pleasant smile melted: "Oh my God. Don't say that." "Look, in my mind this *is* anthrax until proven otherwise. But we don't know yet, so don't say anything to anyone."

Kandy was so frozen by the thought that for days she refused to talk about it. Even her family didn't find out about her involvement until the news became public. She calls the experience "exciting and scary, like nothing else in my life before." And, "I hope I never see anything like it again."

Physicians are largely unfamiliar with many of the germs likely to be adapted as biological weapons. The *variola* virus, which causes smallpox, killed hundreds of millions of people in past centuries. But by the mid-20th century, vaccinations against the disease had markedly reduced the numbers. The last recorded case in the United States was in 1949. A global vaccination program by the World Health Organization to eradicate smallpox from the earth was declared successful in 1980. Thus, few doctors practicing today, especially in the United States, have ever seen a case of smallpox.

Plague is equally obscure. Caused by the bacterium *Yersinia pestis*, outbreaks in the past wiped out huge populations. In the Middle Ages a plague epidemic killed one-third of Europe's population in just four years, 1346 to 1350. The disease became less fearsome after the development of penicillin and other antibiotics in the mid-20th century, since they offer protection if administered soon after exposure. In the United States, beyond a few locations mainly in the Southwest, where an infected rodent occasionally transmits the bacterium to a human, today's physicians have never treated a case. Timely diagnosis thus is less likely.

Still, in the cases of smallpox and plague, historical experience provides a bank of knowledge. Symptoms, treatment, and methods of prevention are well established. None of this is true for anthrax, which in recent centuries has never been widespread among humans. In the United States only 18 cases of anthrax from inhaled spores were recorded in the 20th century. Dr. Bush's early suspicions therefore seem all the more remarkable. Despite the increased publicity about bioterrorism, the cold fact is that the use of biological agents for hostile purposes has been rare. The only known large-scale incident in the United States was in 1984, when the Rajneesh cult in Oregon poisoned restaurant salad bars with *salmonella* bacteria. At least 750 people became ill, but none died. As far as we know, anthrax had never been used in this country for hostile purposes—never deliberately to infect or kill anyone. Dr. Bush was running against the grain of history and experience.

Anthrax spores normally lie beneath the surface of the soil. Grazing animals, like sheep, goats, or cattle, may become infected by ingesting or breathing in the bacteria. Human anthrax infections almost always arise from contact with such an infected animal or its wool, hair, or hide. Cutaneous anthrax, which occurs if spores enter through cuts or other skin openings, is largely treatable with antibiotics. But if the spores are inhaled, they are far more likely to be deadly. Unless antibiotics are administered soon after exposure, recovery is uncertain. Moreover, 90 percent of *untreated* victims of inhalation anthrax die.

In spore form, anthrax bacteria are tough and durable. Potentially dangerous anthrax spores have been found in locations where infected cattle carcasses were buried 140 years earlier. After lying dormant for decades, certain conditions can transform spores into active, germinating organisms. Paradoxically, one such condition occurs if a spore is engulfed by a macrophage, one of the body's natural defense cells that ordinarily destroy such foreign bodies. Thus, the very cell that usually protects a person from an invading microorganism may transport and activate an anthrax spore into a germinating organism that reproduces and releases deadly toxin. It is the durability and lethal power of the spore that make anthrax an attractive biological weapon. Dr. Bush's presumption that someone had been deliberately infected with anthrax took matters across a divide with immense and frightening implications.

Soon after Bush settled on the possibility of anthrax, he had a sample of the bacteria delivered to Anne Beall, the head medical technologist at Integrated Regional Laboratories in Fort Lauderdale, Florida. The lab, which is used by several hospitals in the area, performs tests in addition to those done in the hospitals. One test determines the germs' motility—whether they are capable of spontaneous movement. Motility can be observed through the microscope or sometimes through the spread of growth in a culture medium. Another test involves the action of the bacteria on red blood cells. When placed in a medium containing red cells, certain bacteria destroy the cells, a process called hemolysis. Completing these assessments takes time, maybe 6 to 12 hours, because quantities of bacteria must first be grown and then applied to the tests.

The regional lab received the sample at about 9 a.m. Bush called over to the lab: "Here's the deal, Anne. I think this could be anthrax. We've got to do these tests now so that we can see if we should move ahead with this." Beall began setting up the tests immediately. By early afternoon she had results. The bacteria were nonmotile and nonhemolytic. *Bacillus anthracis* is nonmotile and nonhemolytic. "You know what?" he told her, "these tests don't fit with the other bacilli I thought could be alternative choices. This is really looking quite like anthrax."

Bush then called the state laboratory in Jacksonville and shared his suspicions with the head microbiology technician, Philip Lee, who was also Florida's biological defense technical coordinator. "Phil, we're going to overnight the organism to you. What can you do up there that will give us fast results?" Thanks to the national Centers for Disease Control and Prevention, Lee's lab was well positioned to identify the organism. Based in Atlanta, the CDC is the lead federal agency charged with protecting the health and safety of the American public. Part of the U.S. Department of Health and Human Services, its activities range from monitoring disease outbreaks to providing citizens with health information, from developing vaccination programs to performing advanced laboratory testing to identify microorganisms. In 1999 the CDC established the bioterrorism preparedness and response program to help state laboratories adopt uniform testing methods for suspected bioagents. Dubbed the "laboratory response network," by 2001 about 80 labs around the country were participating.

Not only was the Florida state laboratory in Jacksonville part of the network, but Lee himself had taken courses at the CDC on how to test for suspected bioweapons, including anthrax. "We'll be doing three tests," he told Bush. "What we'll be looking at is capsular staining, a polysaccharide test of the cell wall, and something called a gamma phage test."

The first test, capsular staining, identifies whether the bacillus has a capsule, a thick outer coating. The capsule exists only when the organism is in vegetative form, not in spore form. In its vegetative form the organism reproduces and releases toxin, whereas in spore form it remains dormant and durable for an indefinite period. Through chemical and temperature manipulation in the lab, anthrax bacteria can be induced into either spore or vegetative form. The capsular test by itself is not conclusive for anthrax because some other bacilli are also encased in capsules.

The second test, the polysaccharide test, is performed on the cell wall of the organism when it is in spore form. The wall contains a specific sugar, a polysaccharide that is peculiar to anthrax and a few other bacillus species. As with the first test, a positive result does not mean confirmation. But if both the capsule and the polysaccharide tests are positive, the bug is almost certainly anthrax.

The third test, the gamma phage test, is based on the fact that certain viruses, called phages, can enter and infect bacteria. Among a group of them known as gamma phages, one type can uniquely infect anthrax bacteria. Once inside the bacteria, the phages rapidly reproduce and cause the bacterial cell to split open. The test involves introducing these gamma phages into a population of bacteria. If the bacteria break open, or lyse, they are virtually certain to be anthrax.

By Tuesday evening the specimens were en route to Jacksonville, though Lee would not receive them until noon the next day. Before speaking with Lee, Bush had placed another important call. Around 2:30 he rang up Jean Malecki, the Palm Beach County health director, but she was not in her office. JFK Medical Center is in the county's jurisdiction, and she and Bush had a close working relationship. Ironically, on that day, Dr. Malecki was attending a conference that she had organized on chemical and biological terrorism. Eighty physicians and other healthcare professionals were learning about what to do in case of a bioweapons attack. Half an hour after Bush called, Malecki's assistant handed her a list of messages, including the one from Dr. Bush. She stepped out of the conference hall to a phone in a nearby office and dialed his number:

"Hi Larry. What's up?"

"Jean, just a minute—I'm going to close my door. Are you by yourself?"

"Yes. OK, I'm closing my door, too."

"Look, I've got this guy here, and his clinical course is unusual." Bush summarized what he had found with Bob Stevens and then said, "I think he could have anthrax."

"No. You really think so?"

"I can't prove it, and there is no reason to believe it's based on anything he's been exposed to, but otherwise it sure seems to fit. So yeah, I think he may well."

Dr. Malecki, a tall woman with flowing red curls, has presided over local responses to hurricanes, tornadoes, food poisoning, West Nile fever, and Legionnaire's disease. She joined the county health department in 1983 and became its director in 1991. Malecki revels in her work, which, she announces with pride, is to oversee the largest health department in the state. She emphasizes the difference between her work and that of state and federal public health agencies:

> Tragedies always happen at the local level. People come in to help from the state and federal levels, and then they leave. The tragedy continues here, and you get to know the families. You get to see the post traumatic stress syndrome. You get to have people cry in your arms and take care of their babies. It's very human on the local level.

Jean Malecki and Larry Bush were a perfect match for the needs of the moment. Both were quick to assert their convictions and challenge conventional thinking. They discussed which organisms, besides anthrax, might be causing Bob Stevens's symptoms. They mapped a course of action pending results from the Jacksonville lab. Above all, they agreed to keep the matter quiet, though Malecki said she would call the state epidemiologist at the Department of Health in Tallahassee, Steven Wiersma. Wiersma, she was sure, would notify the CDC. But they worried that premature leaks might cause panic among the hospital staff and patients.

Dr. Wiersma has a vivid memory of the call he received from Jean Malecki that Tuesday afternoon: "We never used the word 'anthrax,' but we both knew what we were talking about. I keep thinking back to that and how peculiar it was that we never said the word." In fact, Wiersma was initially dismissive:

> Jean said that she heard from a clinician, Dr. Larry Bush, who had seen a patient who is critically ill, has meningitis, and he claims there's a Gram-positive bacillus. I mean, we get calls like this, and I think, "Oh yeah, right." So I think, "There are plenty of alternative expla-

nations." So I say, "Are you sure this is Gram-positive and not Gram-negative *diplococci*?" You know, that's the usual call we get on meningitis cases. And I got Jean to waver a little bit. I think she may have started to question herself; in fact, at one time she said, "No, it was Gram negative." So there was some confusion.

I mean that was okay with me. I could put this back into my comfort level and think, "Well, this is probably *Neisseria meningitidis*, a Gram-negative bacterium associated with meningitis, which we get called on a lot. I think I maybe purposely confused her because that's kind of what I do. You know, I try to put holes in people's stories. Of course, even if it was a Gram-positive bacillus, it didn't necessarily mean anthrax. But that's why she called me with such urgency—because she was thinking this could potentially be anthrax. But see, no one mentioned the word "anthrax." She didn't mention the word to me, and I didn't mention it to her.

Dr. Wiersma says he doesn't know why they never said the word, but the symbolism seems obvious. Trying to "put holes" in a claim of anthrax would be a perfectly appropriate role for the state's chief epidemiologist. Moreover, his seeking a "comfort level" with an alternative diagnosis seems understandable. But failure to utter the word reaches beyond science and medicine to raw fear and avoidance. Wiersma and Malecki were engaged in a peculiar dance of denial. Mentioning "anthrax" would somehow make the worst possibility more likely—that the bug actually was *Bacillus anthracis*. Their avoidance is a measure of the enormous significance they knew such a diagnosis would have. It is also clear that while Jean Malecki may have wavered, as Wiersma suggests, she was more inclined to believe the worst than Steve Wiersma was. Her recollection of conversations with him and with other state and federal officials was unadorned: "When I said 'I think we may have anthrax here,' nobody believed me."

Dr. Wiersma decided to speak directly with Dr. Bush and "kind of play the same game I did with Jean":

So I call Dr. Bush and basically play dumb with him. I say, "Dr. Malecki just called and I understand you've got this patient. What do you think this is?" He says, "Meningitis." I say, "Great. You're getting Gram-negative *diplococci*?" He says, "No, no. Gram positive." Then I went through some other Gram-positive species, and I say, "Do you think this is maybe strep pneumonia?" He responds very strongly, "No, no, not strep pneumonia." The way he said it sounded as if he meant, "You idiot."

Dr. Wiersma chuckled as he described Bush's reaction. "But that was my purpose. I was testing him. By the end of the call, it was clear this was something unusual. It was something he had not seen before, and further work needed to be done—looking at the organism." By then the organism was on the way to the state lab in Jacksonville. Wiersma eagerly looked forward to confirmation that the bug was not anthrax. Bush and Malecki eagerly awaited confirmation that it was.

The next day, Wednesday, Bush called Phil Lee. The microbiology lab technician told Bush that the organisms had just arrived and that he would have results of two of the tests by midafternoon. The gamma-phage test could not be completed until the next morning. Later in the afternoon, Lee called Bush and told him that the capsular test was positive—the organism had a capsule. But the second test result, for polysaccharide in the cell wall, was equivocal.

"How soon can you repeat the cell wall test?" Bush asked.

"It takes a couple of hours," Lee said.

At day's end they spoke again: "I'm still getting an equivocal response," Lee said.

"Are you willing to call this *Bacillus anthracis* based on what we have?" Bush asked.

"No, not at this time. I can't."

Jean Malecki was also in touch with Lee and was hearing the same message. Lee promised both of them a gamma-phage result the next morning, probably by 8:30. But she was already so convinced of the probability that she decided that afternoon to open a formal investigation. Meanwhile, Bob Stevens's condition was worsening, now showing signs of kidney failure. He was unlikely to regain consciousness soon, if ever, so Malecki could not expect information from him. That Wednesday she and five members of her department practically spent the night at the hospital. They reviewed Stevens's medical records from prior years, spoke to doctors and hospital staff, and interviewed his wife, Maureen, intermittently through the evening, until 2 a.m.

Malecki's report cites all of Bob's recent activities outside of work, to the extent that Maureen could recall them. References go back to June, when they visited England and "had taken a couple of walks in the country." A daily summary begins with an entry for Saturday, September 22, five days before they left for North Carolina. On that day Robert Stevens:

> Possibly went fishing (place unknown). His usual fishing spots include the pier at Boynton Beach Inlet (saltwater, ocean fishing); Loxahatchee Wildlife Preserve (fresh water); and also Lake Osborne in Lake Worth. He does not use a boat; he fishes from the banks. He does not hunt.
>
> Out to dinner with friends at Roadhouse Grill in Boynton Beach. He had steak, a small sirloin.

The entry for the next day, Sunday, September 23, reads: "6:30 to 8:20 a.m. Bicycling with friend at John Prince Park, Lake Worth. Later to Lake Worth Beach with 11-year-old granddaughter to possibly the public pool or arcade there."

While not mentioning anthrax, the report pointedly cites items that might have carried spores. Stevens visited the Fortune Cookie Oriental Store, where "they have fresh meats, including goat meat" and an Indian store and restaurant in West Palm Beach—"there is also goat meat there." He recently bought "leather shoes at the Rack Room in the Boynton Beach Mall." Another entry notes: "Within past month he pulled weeds out of an overgrown area where he keeps herbs in pots."

None of these activities or visits explicitly suggested a connection to terrorism. But Dr. Malecki, like Dr. Bush, doubted that the anthrax bacteria that had presumptively infected Mr. Stevens would have come from a naturally occurring source. (Their suspicions arose from their knowledge that inhalation anthrax in the United States is extremely rare.) In any case, even their belief that Mr. Stevens had anthrax was still not widely shared by state or federal officials.

What neither Bush nor Malecki knew was that after Philip Lee's second equivocal cell wall result, he did the test again. But for this one he decided to grow more vegetative cells before testing them. He placed the bacteria in a nutrient mix and incubated them for one hour at 37° Celsius (98.6° Fahrenheit). More cells grew, with more cell walls to test for polysaccharide. Shortly after 10 p.m. on

Tuesday evening Lee had a result. This time it was unambiguously positive.

Lee immediately called the CDC in Atlanta and left a message that he needed to speak with a senior official. The agency had been aware of Bush's and Malecki's concerns since Tuesday when notified by Steve Wiersma, the state epidemiologist. On Wednesday CDC officials repeatedly spoke with Wiersma and then with Phil Lee to keep abreast of his lab findings. Like Wiersma, the CDC people were skeptical. Since 1998 they had analyzed hundreds of materials and specimens ostensibly containing anthrax. Many, as cited in press reports, were powders accompanied by letters claiming that the reader had been exposed to anthrax. All had proved false. Accordingly, CDC officials were not about to jump to a hasty conclusion. But they did request a sample of the material that was being tested in Florida, which Lee sent out earlier that afternoon. The package was expected to arrive the next day, Thursday.

By 10:30 Wednesday evening, Lee had spoken to several CDC people about his positive cell wall test, and they all decided to wait for the gamma-phage results due the next morning. An hour later, at the end of a 15-hour day, Phil Lee left for home. He was tired and anxious about what the results would look like the next day. Actually, the test was already under way. He had inoculated a culture of the bacteria with gamma phage and placed it in an incubator. The mystery remained whether that night the viruses would invade the bacterial cells, feast off them, multiply, and cause them to burst.

On Thursday, October 4, at 8:15 a.m., Lee donned his gloves, protective outerwear, and clear plastic face hood and prepared to bring the culture into the isolation lab. The gamma phages had now lain with the bacteria for some 12 hours. He set the culture plate under the microscope and stared. "When I saw it, my heart rate went up." Beneath his eye lay fields of decimated bacteria exploded by battalions of viral parasites. That moment Lee knew with certainty that the bacteria cultured from Robert Stevens were anthrax. "All I had going through my mind was the hope that this was a natural case rather than a terrorism incident." But time, circumstance, and his own training left him doubtful that the cause was not man-made. He took off his laboratory outerwear and headed back to his office to call state and CDC officials. As he walked through his office door, the phone rang.

"Phil, this is Larry Bush. How's your gamma-phage test?

"It's positive."

"OK, now are you going to call it anthrax?"

"Yeah," Lee said, uncomfortably. Then he added, "I think I should have told the state people and Dr. Malecki before I told you."

Bush, characteristically impatient with bureaucratic niceties, answered, "What's the difference? So you're going to hang up the phone from me and call them. Look, I'm the treating physician. I sent you the organism. You had the obligation to tell me." Bush later offered a more sympathetic assessment:

> I guess he felt he should call the people he had worked for first. I thought it was odd, but of course the implications are bioterrorism. Here's a guy who recently comes from CDC training to set up their lab training-and-response network and gets an organism they are set up to look for. What that means in his mind, and obviously in mine, is bioterrorism. So we get off the phone. I call Jean Malecki. He calls the CDC. I tell Jean. She calls him. Within 10 minutes we all had the same information.

Phil Lee's worry that he had violated protocol was short lived. The enormous implications of the findings quickly overtook such mundane concerns.

While Lee and Bush were on the phone with each other, Jean Malecki was in the middle of a conference call with Steve Wiersma and other state officials. They were reviewing what was known so far. By now she was convinced that Bob Stevens had anthrax, though others on the call remained doubtful. Malecki recalls Wiersma saying, "We still have to wait. It may be negative." Another official on the call agreed: "Well, you know, it's not in his lungs, nothing respiratory." Malecki answered: "You can still have 50 percent of these cases present this way." "OK, but there are other organisms," the official insisted. Malecki thought to herself, "But I don't know which other ones they could be at this stage of the game."

Just then, Malecki's assistant signaled her that Phil Lee from the Jacksonville lab was on the other line. She excused herself to take his call. When she returned to the conference call, she mentioned that she had just heard from Lee and dryly added, "Well, it's positive, Steve." Wiersma responded, "I gotta get off the phone. I gotta get down there."

The implications of the anthrax confirmation were not lost on anyone. Calls went quickly back and forth among state, federal, and local officials. Dr. Tanja Popovic, a leading laboratory investigator at the CDC, had trained Lee to do the tests and felt a personal sense of satisfaction about his performance. "Philip," she told him, "your testing was perfectly accurate."

The Florida Department of Law Enforcement arranged to fly Wiersma and another health officer from Tallahassee to Palm Beach. They arrived about 11 a.m., rented a car, and drove to Jean Malecki's office. By then the CDC in Atlanta was also organizing a team to fly down later in the afternoon.

As soon as Wiersma arrived at the Palm Beach County office, he and Malecki reviewed copies of Bob Stevens's medical records. After that he asked members of his staff to start looking for additional cases and to "design a surveillance strategy." He went to the hospital to meet Dr. Bush and to talk to Maureen Stevens and other family members, including Casey, who by then were at Bob's bedside. Later, after leaving the hospital, Wiersma was just as puzzled as before about how Stevens could have become infected.

Still later in the afternoon, the CDC team arrived and met with Wiersma's and Malecki's staff in a conference room down the hall from Malecki's office. Dr. Bradley Perkins led the CDC group. His position as chief of CDC's branch on meningitis and special pathogens was aptly named for this mission. Mr. Stevens had meningitis, and the bug they were dealing with, *Bacillus anthracis*, was surely special. "We all agreed on two priorities," he recalls. "First, to determine how the exposure occurred, and second to identify other possible cases." They organized investigation teams each with representatives from the three agencies—federal, state, and local. On Friday morning the teams began to sweep through the places where Stevens lived, worked, ate, shopped, fished, hiked, biked, and visited. The CDC had also dispatched a group to cover his trail in North Carolina the week before. Everywhere they went the public health teams sought leads about how Stevens became infected. They interviewed people, swabbed surfaces, and collected samples for testing.

Previously, the last recorded domestic case of inhalation anthrax had been in 1976, when a California weaver was infected by spores in yarn made from Pakistani wool. Still, officials publicly minimized the possibility of a deliberate attack. "We have no reason to believe at this time this was an attack at all," Steve Wiersma told the *Palm Beach Post* that Friday. "There is no evidence of terrorism," announced U.S. Health and Human Services Secretary Tommy Thompson from Washington. "It appears that this is just an isolated case." He mentioned that Stevens "drank water out of a stream when he was traveling through North Carolina." Speaking about the incident, coming so soon as it had after September 11, Florida Governor Jeb Bush was emphatic: "People don't have any reason to be concerned," he told reporters. "This is a cruel coincidence. That's all it is."

Brad Perkins was determined to try to find a natural cause. He acknowledged that such a case so soon after 9/11 and in a state where the hijackers had lived and trained at flight school was "slightly chilling." But he would not let that influence the manner of the search:

> One of the risks that I wanted to control was making sure that we didn't miss some natural exposure. It would have been easy to leap to the conclusion that this was bioterrorism and the result of an intentional release. But I thought one of our important roles as scientists was to not miss a natural exposure and this being a chance occurrence.

The fact that agents from the Federal Bureau of Investigation were part of the investigation teams was downplayed to the public. The agents were going along "just in case anything is found," said Judy Orihuela, a spokeswoman for the bureau.

Brad Perkins was on the team investigating Stevens's residence and workplace. When they got to his house, Perkins and the other team members pulled on latex gloves and swiped surfaces with moist sterile cotton swabs. They then inserted each swab into an individual container marked with the time and location. At the house they took samples from indoors and outdoors—swipes from the kitchen counter and the bathroom sink, yard soil, snippets from the small vegetable garden. The samples were delivered to the state laboratory in Miami for analysis. Miami was closer than Jacksonville, only an hour south of the investigation area on Interstate 95. The lab there could also accommodate the beefed-up numbers of technicians that CDC was sending down to facilitate testing.

Before starting the field tests on Friday morning, Perkins stopped at the hospital to obtain permission from Maureen Stevens to enter her house. She impressed him deeply: "I mean, given the situation, she was extremely gracious and composed." Her demeanor was all the more impressive considering what she was enduring beyond having a gravely ill husband. By then word about her husband's diagnosis had been flashed around the world.

Larry Bush had not shared his early suspicions with Mrs. Stevens. She didn't learn until Thursday that Bob had anthrax, after confirmation from the state laboratory. But as early as Tuesday, Bush's questions were suggestive. Not that Maureen was likely to understand their implications. Those could only have occurred to someone unusually attentive to the occasional news stories about anthrax threats and hoaxes sent through the mail. Dr. Bush, however, *was* aware of them, which helped him frame his questions:

> Beginning Tuesday, as we get more information about the possibility of anthrax, I'm going back to her wondering, Why would this guy have anthrax? He was a photo editor for the *Sun*. I mean the AMI building is the publication office for all those tabloids. I kept saying to her, "What exactly does he do?" She said, "He looks at photographs that come to him from all over the world, and he edits them for what goes into these tabloids." I asked, "So how does he get these pictures?" "They come over the Internet."
>
> I asked, "Do they all come over the Internet? How else does he get them?"
>
> "Well, he gets some by mail."
>
> I asked, "Does he open the mail?"
>
> "Yes, he opens the mail."
>
> "Did he ever tell you anything unusual happened with the mail?"
>
> "No, nothing."
>
> She didn't know what I was looking for. I mean she knew that I was looking for a cause of his overwhelming infection, but I never mentioned the word "anthrax" to her. I believed he had anthrax, and in my mind it was bioterrorism. He didn't get anthrax from being in North Carolina. He didn't get anthrax from being a fisherman. There were big reports that he was an outdoor person. I mean we're in Florida. Everybody here is an outdoor person. Everybody fishes.

On Thursday around 1 p.m., Maureen Stevens was ushered into a hospital conference room where for the first time she was told that her husband had anthrax. Dr. Bush was present along with Steve Wiersma, hospital staff, and an FBI agent. Bush sensed she was disappointed that he hadn't discussed the possibility with her earlier. He thought to himself about the past 3 days, what he could have said, when he could have said it. He just wasn't sure. "It was hard," he recalled. "Truthfully, I didn't use the word 'anthrax' because I didn't think it would mean much to her." He was also afraid that if word got out it could cause "a whole public event here at the hospital."

An hour after Maureen learned that her husband had anthrax, there was indeed a public event. The hospital held a press conference, and Bush felt even sorrier for her.

> All of a sudden, this overwhelming event takes place—with the FBI, the CDC, the press, and everyone being there. And we're going in front of the microphones saying the man has anthrax. She had been asking me, "Does he have any chance of survival?" —before we used the word "anthrax." And I'm saying, "Everybody who is still alive, obviously, with any infection has a chance of survival. But he is critically ill."
>
> Then we're at this press conference with all the news media there, and they're asking, "Is he going to live?" and this and that. And she's sitting in the front row hearing all this. It was overwhelming for her. She goes from hearing her husband is ill with meningitis to being in a huge conference room with the world's press. Out in the parking lot there must have been 50 trucks with satellite antennas.

If Maureen felt overwhelmed at the press conference, it was worse when she went home that evening. By then Jean Malecki had spent many hours with Maureen. "She is a beautiful woman," Malecki says, "and I became very close to her and the family." Dr. Malecki sounded pained when she recalled what happened after the press conference:

> I felt so sorry for that family. It only takes a second for the press to find out who the person is, even though public health officials never reveal it. Literally, when this thing hit, they surrounded her home. They were paying $50 a night to spend the night on roofs of her neighbors' homes. These big trucks that come along with the TV media, like CNN, surrounded her house. They were on her roof and everywhere. She had to put up blankets to block the lights so she could get some sleep. And then she had to get security people to guard her home.

Public health officials do not normally cite patients by name out of respect for their personal and family privacy. Months after Bob Stevens's illness, CDC and state health officials continued to refer to him as an "anthrax case" or, because he was the first, "the index case." But for Jean Malecki, once a patient's name has been broadcast widely and is part of the public conversation, it seems cold and unfeeling to maintain that convention. "It's true that CDC people and others are trained to do that," Malecki says, "but I don't." She underscored her point by affirming Bob's humanity:

> I came to know Mr. Stevens's family very well. His name was Mr. Robert Stevens, a 63-year-old English gentleman who migrated here and had a beautiful family. He was an earthy kind of guy, an environmentalist sort, and tragically died of anthrax. I give lectures all over the country, and I use his full name.

That Friday afternoon, October 5, after investigators finished at the Stevens's home, they went to his place of work. In the midst of taking samples at American Media, Inc., Perkins's cell phone rang. The caller informed him that Bob Stevens had just died. Perkins was taken aback. He had seen Stevens hours earlier and though he knew Stevens was critically ill, he had seen no indications that death was imminent. Perkins told the rest of the team, and after a brief pause they resumed swabbing and sampling. Their quest for answers seemed to have become more urgent.

By Saturday the Miami laboratory was processing samples from North Carolina, from the docks where Stevens fished, the park where he biked, and from dozens of other spots, including his home and the AMI building. The next day, Sunday, preliminary testing indicated that *Bacillus anthracis* was present in samples taken from the AMI building mailroom and from Bob Stevens's computer keyboard. The results would not be confirmed until additional testing was completed. But Perkins had shed his agnosticism and become a believer. "There was no reason that *Bacillus anthracis* should ever be in this workplace," he thought to himself. "This was an intentional exposure."

The discovery of anthrax at AMI raised enormously important questions: How did anthrax get into the building? Was anyone else there infected? Were others at risk? Was the AMI building the sole target? Who was responsible for releasing the germs and why did he, she, or they do it?

CHAPTER
TWO

AMERICAN MEDIA

On Monday, October 1, 2001, the day before Bob Stevens entered the JFK Medical Center in Atlantis, Florida, Ernesto Blanco was admitted to Miami's Cedars Medical Center, 65 miles to the south. "Ernie," the name favored by his fellow American Media, Inc., employees, was the company's mailroom clerk. At 73 he had an elfish smile and a full head of hair, more black than gray, which suggested a man 10 years younger.

Toward the end of the previous week Ernie had been "feeling something funny," he recalls in accented English. The Spanish-language rhythms of his native Cuba, where he had been an accountant before fleeing the 2-year-old Castro regime in 1961, still flavor his speech. On Friday fellow workers noticed that Ernie seemed dazed. As noon approached, Maria Wolcott, a clerk in the advertising department, called the security office to see if someone there might drive him home. "Maria, why did you do that? I feel fine," he protested. His bravado was short lived. Soon after, he confessed to feeling dizzy to his boss, Daniel Rotstein, AMI's vice president for human resources and administration. "Go home and don't worry," Rotstein said.

After retiring from his own carpet installation business in 1989, Ernie began working for Globe Communications, a tabloid publisher. Ten years later AMI acquired Globe and then moved its headquarters from Lantana to the Globe building in Boca Raton, where Ernie continued to work. His daily commute from his Miami home began with a 7-minute drive to the Golden Glade Tri-Rail station. After parking, he'd board the 6:35 a.m. train north. Sixty minutes and seven stops later he was at the Boca Raton station, a short bus ride away from the AMI building on Broken Sound Boulevard.

By 8 a.m. Ernie was driving the company van to the post office to pick up the mail, somewhere between 3,000 and 5,000 pieces a day. Upon returning to AMI, he would wheel the pile into the mailroom. Then began the sorting process, which he usually did alone, though occasionally with help from Stephanie Dailey or another clerk. With lightning speed, Ernie flipped letters into their mailboxes, a honeycomb of 300 open compartments. "I know where most of the letters go without looking at the names on the mailboxes," he says. No small achievement, considering the plethora of departments, offices, and individual addressees at AMI. Once sorted, the letters were transferred to mail carts for delivery throughout the three-story building.

In the course of wheeling the carts around, Ernie came to know people everywhere. On the ground floor, near the mailroom, was Martha Moffett, who ran the library. Down the hall were the offices of the legal staff, the photo library, and the personnel department. An elevator brought Ernie to the second floor, where the *Globe*, the *Star*, and the *Weekly World News* had their offices. On the third floor, continuing his deliveries, he'd wave to Ed Sigall, senior editor of the *National Enquirer*, and to Ray Villwock, editor-in-chief of the *National Examiner*. If Ernie saw Bob Stevens at the *Sun*, they were sure to exchange a quip or two. In the sweeping executive offices next to the *Sun*, David Pecker, AMI's president and owner would offer a smile. Ed Sigall thought of Ernie as a company fixture whom "everyone knows and likes."

On the Friday he felt dizzy, when the head of security came to take him home, all Ernie remembers hearing was, "Okay, Ernie. Let's go home. I'll drive." An hour later he was back in the Miami house in which he had lived since 1976. His wife, Elda Rosa, dosed him with hot lemonade, honey, and aspirin. He felt a bit better. Convinced that he had the flu, Ernie was sure he'd be back at work on Monday. But as the weekend progressed, he developed a cough and fever, his fatigue worsened, and he became more disoriented. Early Monday morning Elda drove him to the Cedars emergency room.

Ernie was admitted and diagnosed with pneumonia. He was given intravenous azithromycin and cefotaxime, antibiotics that are effective against a range of bacteria. But Ernie's delirium worsened and he began to have difficulty breathing. On Tuesday the treating physician called in Dr. Carlos Omenaca for consultation. Omenaca, tall and dark, as handsome as any character on TV's "General Hospital," speaks with a light Castilian accent. In the 1980s, after attending medical school in his native Spain, he came to New York to do research at New York's Hospital for Joint Diseases. He studied further at New York University and completed a residency in infectious diseases and critical care at St. Luke's-Roosevelt Hospital. In 1997 he moved to Florida.

Omenaca examined Ernie, looked at his chest X ray, and concurred that he had pneumonia. During the next few days, Ernie failed to improve. His X ray showed a pleural effusion, a collection of fluid in the space between the lungs, which suggested bacterial pneumonia. But Omenaca says, "The funny thing was that he had a typical presentation of a viral illness. There was nothing else to suggest bacterial pneumonia." Omenaca was referring to the fact that Ernie's fever, nausea, confusion, and fatigue were consistent with viral symptoms. At the same time, his blood culture, taken the day after his admission to the hospital, revealed no suspicious bacteria. If the usual bacterial suspects for pneumonia, such as *pneumococci* or *staphylococci*, were responsible, a day or two of antibiotics should have produced some improvement. By Thursday Omenaca felt stymied:

> We had X rays consistent with bacterial pneumonia in someone who did not have the clinical picture of it. So you have to start working up all the atypical pathogens. Ernie had five or six dogs at home, so I thought maybe he has leptospirosis [an infection that can be transmitted by rats, dogs, or cattle] or something from the animals. It was a very bizarre presentation.

On Thursday morning, October 4, Daniel Rotstein was in his AMI office when he heard from the Palm Beach County Health Department that Bob Stevens, whom he knew was in the hospital, had meningitis. Worried about the effect of this news on other employees, Rotstein contacted a medical consultant the company had used in the past: "We need to draft something about meningitis to tell the employees what they need to do and not do." He also was considering inviting a doctor in to answer questions the employees might have. Later, in the afternoon, his secretary told him that Bob

Stevens's wife, Maureen, was on the phone. He was stunned to hear her say, "I just found out Bob has anthrax." "We'll do all we can to help," Rotstein assured her. After hanging up, he realized that all he knew about anthrax was that it was scary.

Soft spoken and professorial, Rotstein, 36, confided, "You know, the first thing I did after that was call my father." Was he a frightened young man seeking parental comfort? Not really, he grins. "My father is a physician, and I asked him about anthrax." Like most doctors, Rotstein's father, a neurologist and psychiatrist, had limited knowledge of the disease. "He told me what he could. He also knew about that island in Scotland."

Rotstein was referring to Gruinard Island, a small body of land, less than 2 miles off the northwest coast of Scotland. His father knew that the island had been contaminated with anthrax, but little more. In fact, during World War II, British and American scientists had performed biological warfare experiments there. Seeking to determine how lethal anthrax spores would be after they were released by explosive devices, the scientists conducted several tests. In one they tethered sheep at measured distances from a central point. Then they exploded cannisters that had been filled with a gruel containing anthrax spores.

Within a day the animals were dead or dying. After more experiments, anthrax-infected carcasses began to litter the site. The scientists and their staffs removed or incinerated the remains. In 1943, when the testing program ended, they set fire to the island, anticipating that the spores would be destroyed. But by the end of the war the scientists were surprised to find that spore counts in the soil were undiminished. In succeeding decades, periodic testing confirmed that concentrations of anthrax remained unabated. For more than 40 years, no one was permitted to go on the island. Not until 1990, after soaking portions of the island in tons of formaldehyde and seawater, did the British government deem it safe for human visitation. By then Gruinard Island had become a monument to the danger and durability of anthrax spores.

Soon after Maureen Stevens called Rotstein, news of her husband's anthrax "started hitting the media," as Rotstein put it. Within minutes of the first broadcast, he received a call from Martha Moffett, AMI's head librarian. She was home with a stomachache and diarrhea. "I had just taken my temperature and was looking at the thermometer when I heard a radio news bulletin," she recalled. "It said that an employee at AMI had been diagnosed with anthrax."

Moffett, who holds a library science degree from Columbia University, began working for the *National Enquirer* in 1976. She had been thinking about retirement, but AMI management convinced her to stay on. Now, alone in her small Lake Worth home, surrounded by shelves of books and papers, she spoke nervously into the phone:

"Dan, I just heard the radio announcement. Look, I'm sick and I don't know what it is."

Rotstein replied, "We really don't have much information about all this yet."

"What should I do?"

"Go to the hospital immediately and tell them what you just told me."

Rotstein said he believes in "erring on the side of caution." Moffet packed a few things, drove to Congress Avenue and headed south less than a mile to the nearest hospital—JFK Medical Center, in Atlantis. "I knew Bob Stevens was there," she said later, "but it's my hospital. That's why I went there."

The other AMI employees also quickly learned about the anthrax "because we have CNN on all the time," Rotstein said. It was almost 5 p.m., and people were leaving for home. He heard several express concern about Stevens's condition. They were plainly eager for more information. Rotstein remained at his desk, planning for the next day. He thought about his earlier idea to notify everyone about Stevens's meningitis. Now with anthrax in the picture, a meeting the next day with all employees seemed imperative.

Sometime after 7 p.m., as he was developing an agenda for the meeting, a frightening thought popped into Daniel Rotstein's head. "I put two and two together," he said. "I remembered that when Ernie Blanco left here, he was very disoriented." People from AMI had kept in touch with Ernie's family, and Rotstein knew that Ernie had been diagnosed with pneumonia. But to Rotstein the symptoms sounded similar to those that had been reported of Bob Stevens. "I decided to call his hospital to alert them to the Stevens situation."

Rotstein reached the Cedars operator and asked to speak to Ernesto Blanco's floor nurse. She was unavailable. "Please have her call me as soon as possible," Rotstein said. A half hour passed with no word from the hospital, so he called again. Again he was told that no one involved with Mr. Blanco's case could take his call. Rotstein went home and continued to phone the hospital, each time requesting a return call. After a half dozen futile attempts, at 11:30 p.m. he received a call from the hospital's infectious diseases nurse. She sounded very interested in what he had to say and promised she would give the information to a doctor.

Half an hour later, at midnight, Rotstein's phone rang. It was Dr. Carlos Omenaca. "I told him about the situation with American Media and with Bob Stevens and how some of the symptoms appeared to be very similar," Rotstein later said. Omenaca remembers the conversation with appreciation: "I started looking at Mr. Blanco's case as possible anthrax immediately after he called."

In May 1999 an article appeared in the *Journal of the American Medical Association (JAMA)* entitled "Anthrax as a Biological Weapon: Medical and Public Health Management." Based on a review of the literature on anthrax, the article represented a consensus view of 21 leading medical, public health, military, and emergency management experts. Among them were Donald A. Henderson, Tara O'Toole, and Thomas Inglesby from Johns Hopkins University's civilian biodefense program, and Edward Eitzen and Arthur Friedlander from the U.S. Army's Medical Research Institute of Infectious Diseases at Fort Detrick, Maryland. The experts were described as the Working Group on Civilian Biodefense. Their article was generally viewed as the most up-to-date compendium on anthrax as a weapon.

In treating inhalation anthrax, the working group advised that, initially, a patient be given intravenous doses of ciprofloxacin, or Cipro. Studies had shown that Cipro, more than other antibiotics, was effective against a large variety of anthrax strains. But the article also said that if a strain is shown to be susceptible to doxycycline or penicillin, these antibiotics would be preferable. That is because overuse of Cipro would promote the development of resistant strains of other more common pathogens, including *staphylococci* and *streptococci*. Frequent use could also lead to unwanted reactions such as nausea and diarrhea or central nervous system disturbances and irregular heartbeats.

The working group recommended that patients with "clinically evident" inhalation anthrax receive an intravenous dose of 400 milligrams of Cipro every 12 hours—800 milligrams a day. The amount is consistent with guidelines for treating other infections. The Bayer Corporation, the company that manufactures the drug, recommends daily doses that total 800 milligrams for infections of the urinary tract, respiratory tract, bones, joints, and other locations. For severe and complicated cases, the manufacturer's advisory allows for a daily total of 1,200 milligrams, but no more. A cardinal understanding in pharmacological medicine is that the higher the dose, the greater the risk of adverse reactions.

Over the years Dr. Omenaca had prescribed Cipro many times at the advised dose levels. Like other doctors, he had never seen a case of anthrax, but he knew that Cipro was considered the initial antibiotic of choice. He reacted quickly:

> Immediately after I hung up from his boss, I put Mr. Blanco on intravenous Cipro. He had been on cefotaxime and azithromycin, a typical combination of antibiotics for community-acquired pneumonia. So what I did was place him on IV [intravenous] Cipro, the highest dose I'd ever given.

How much was that "highest dose?" Omenaca smiled faintly, seeming to find incredulous what he was about to say, "Seven hundred and fifty milligrams, three times a day." That's 2,250 milligrams a day! Nearly twice the recommended level for even the most severe infections. Why did he do it?

Omenaca knew that the medical information on human inhalation anthrax exposure was sparse and tentative. As would later become evident, many assumptions about anthrax turned out to be wrong. The available data reflected uncertainty. For example, the *JAMA's* working group article on anthrax cited findings that "the LD50 (lethal dose sufficient to kill 50 percent of persons exposed to it) is 2,500 to 55,000 inhaled anthrax spores." A cluster of 55,000 spores is microscopically tiny—smaller than the eye of an ant. But when compared with the low-end estimate of 2,500 spores, the difference between low and high seems large.

Uncertainty also shadowed information about deaths from anthrax. The handful of known inhalation cases in the United States, just 18 in the 20th century, had a mortality rate of 89 percent. Still, according to the *JAMA* article, "the majority of cases occurred before the development of critical-care units and, in some cases, before the advent of antibiotics." Similarly, in 1979, when germs escaped from a Soviet military facility in Sverdlovsk, 68 of 79 inhalation anthrax victims reportedly died. But here, too, the utility of the data is unclear because, according to the article, "the reliability of the diagnosis in the survivors is questionable."

Dr. Omenaca learned these facts, and more, in succeeding days as he immersed himself in the literature on anthrax. Absent from his readings was any suggestion that Cipro be administered in the amounts Ernie was receiving. Moreover, it was not even clear that Ernie had anthrax: no *Bacillus anthracis* in his blood culture, no widened mediastinum visible on his chest X ray. But every day massive quantities of Cipro nearly three times the recommended dosage—continued to trickle into his system. Dr. Omenaca would not back off.

On Friday, October 5, now aware of Ernie's hospitalization, state and federal public health officials became interested. They arranged for swabs of his nasal passages to be taken and sent for analysis to the Centers for Disease Control and Prevention in Atlanta. During the next few days, Ernie's condition worsened. His blood pressure fell, and, to assist his faltering breathing, he needed high concentrations of oxygen, delivered through a mechanical respirator. On October 7 the report came back from the CDC that one of the swabs had picked up anthrax spores. The finding was tantalizing but not proof that Ernie had anthrax. The presence of spores did not confirm the presence of the disease any more than finding a cold virus in someone necessarily means the person has a cold.

Still, the finding was suggestive. Until then, "my suspicion that it was anthrax was low," Omenaca said. Even if Ernie had the disease, finding anthrax spores at that point was odd. "Usually after 24 hours of antibiotics you get negative testing," Omenaca said. "But he grew out one colony from the four swabs." A colony of anthrax bacteria had grown from one of the samples placed on plates containing blood agar and other nutrient media. The CDC then confirmed, through PCR (polymerase chain reaction) testing, that the bacteria were anthrax.

PCR was developed in 1985 by Kerry Mullis, a scientist then with Cetus, a California biotechnology company. The discovery earned him a Nobel Prize in 1993. Mullis subsequently gave up science in favor of surfing and recreational drugs. But he bequeathed a procedure that has become an essential tool of laboratory science. PCR enables multiple copies of a piece of DNA to be made quickly and repeatedly. In just 1 hour a tiny snippet of DNA that had been exposed to special DNA primers can generate a billion copies. To identify anthrax, a DNA segment known to be part of the anthrax genome is prompted to find a complementary segment of DNA in the suspected organism. If copies are successfully generated, or "amplified," the presence of anthrax DNA in the suspected bug is confirmed. In such a manner the CDC was able to establish that the bacterial sample from Ernesto Blanco contained anthrax DNA.

As of Thursday, October 4, Robert Stevens was known to have contracted anthrax. The diagnosis was confirmed by other tests because the Florida laboratory in Jacksonville was not equipped for PCR testing. Subsequently, the CDC did conduct PCR on Stevens's samples. Not surprisingly, the test was positive for anthrax, just as the PCR on Ernie Blanco's sample was positive. The difference was that the bacteria from Stevens were grown from his blood and cerebrospinal fluid. The spores from Blanco had been obtained by a nasal swab. Again, confirming the *presence* of anthrax spores did not necessarily mean *infection*. Rather, the patient's symptoms and the bacteria's location in the body are key determinants.

Yet another suggestive indicator about Ernie's disease came from CDC testing on the bloody fluid in Ernie's chest. Here, too, PCR testing was positive for anthrax DNA. The anthrax DNA could conceivably have come from an earlier exposure that had never caused disease. But to Dr. Omenaca the test results seemed increasingly persuasive:

> I said, well, he's been exposed to anthrax and he has a bad disease which I don't have an explanation for. Whatever we have in the literature is so scant, it's so minimal that—who knows? Maybe this is an atypical case, or we just don't know enough. So I am assuming that he has inhalation anthrax. So we continued to support him with the intensive care unit and give him antibiotics as needed.

The CDC's refusal to call Ernie's illness "anthrax" frustrated his family. "Dr. Omenaca was telling us one thing, and the Centers for Disease Control were telling us another," said Maria Orth, Ernie's stepdaughter. "They just kept saying pneumonia, pneumonia. And I thought: What? With the security guard posted outside Ernie's room? With the intensive care unit and everything? With the FBI investigating?"

What she did not know was that back in Atlanta, CDC officials were having a running debate. "There were lots of discussions about whether Mr. Blanco was a case or not," recalled Stephen A. Morse, associate director of CDC's program on bioterrorism preparedness and response. "It all hinged on the fact that they could not culture anthrax from him." Would CDC officials remain rigidly attached to their central tenet—the ability to culture, or grow, the bacteria from a patient's sample? In fact, 2 weeks later the CDC announced that Ernie had anthrax. "What happened was they had to change the case definition of anthrax," Morse said. Whether or not a person has a particular disease is determined by specific indicators, including symptoms and test results. With the benefit of experience, the criteria for a determination—or a "definition"—may be revised. Morse explained:

> Part of the case definition had been a positive culture for *Bacillus anthracis*. But Mr. Blanco had received long-term therapy quinolones [Cipro is a fluoroquinolone] and several other antibiotics which killed all the anthrax organisms in his body. So they were not going to get any positive culture. But they were able to get two nonculture tests positive. So they changed the case definition to be the two nonculture tests.

Nonculture tests could include cell wall and capsular staining, PCR, or demonstration of the presence of antibodies to anthrax. If an illness were "clinically compatible" with anthrax but only one test was positive, the CDC would consider the patient a "suspected case."

While CDC was arguing about case definition, Ernie was fighting for his life. "I felt physically, and in my soul, that I was leaving this world," he said later. The family summoned a priest to give him last rites. During the following days Ernie remained conscious but confused. After Daniel Rotstein had called Omenaca, the doctor asked Ernie, "Do you know Mr. Rotstein?"

"No," Ernie replied.

"But you work with him," the doctor said.

"Everybody's asking me that question. I don't know him."

A few days later Omenaca asked Ernie the same question. "Of course I know him," Ernie said. "I know him for 10 years. We see each other every day. We work in the same building."

After 4 or 5 days on the massive doses of Cipro and repeated thorocentesis—drainage of bloody fluid from his chest—Ernie's condition began to improve. On October 23, twenty-two days after he entered the hospital, Ernesto Blanco went home. Weak but on the mend, he continued to take oral Cipro for weeks and to see Dr. Omenaca and other doctors several times a week. In March 2002, Ernie returned to work at AMI on a reduced schedule. Relaxed and happy to tell his story, he said, "Now I feel all right, good. It looks to me I'm perfect."

In the midst of Ernie's darkest days, Omenaca had been in touch with Dr. Aileen Marty. A Navy commander, she is a specialist on emerging infections—previously unknown or rare diseases that are appearing more frequently in the human population. On his behalf, she spoke with several experts on anthrax, including Dr. Arthur Friedlander, one of the author's of the *JAMA* article. She conferred repeatedly with Omenaca, encouraging him about the treatment he was giving Ernie. "Dr. Omenaca is a wonderful doc, a brilliant clinician," she said afterward. "I told him he was doing the right thing by Mr. Blanco."

On the evening of Thursday, October 4, hours after the announcement at JFK Medical Center that a patient there had anthrax, Dr. Larry Bush got a call from the emergency department. By then everyone at the hospital knew that Bush had been the first to suspect that Bob Stevens had anthrax. "We have a woman down here," the voice said. "She has some pneumonia, and she's concerned about anthrax."

"You know what, guys?" he responded. "You can't call me for every cough you get because everybody thinks they have anthrax."

"But that's not it. She really has a story to tell you."

"Okay."

He recounted the experience: "So I go down there. I meet Martha Moffett. She's got a cough. She's got a little fever." He looks at her chest X ray and examines her.

"Do you have a headache?" he asked.

"No."

"Any neck pains?"

"No." Martha said later, "I was scared, but I began to feel better when I could answer 'no' to almost all his questions."

"Martha, I think you have a little pneumonia, but I don't think you have anthrax. They'll give you some antibiotics, and you can go home and follow up with your doctor." Then he added: "I'm curious. Why would you think you got anthrax?"

"Well, because of Bob Stevens. I work with him. I'm the librarian at the AMI building."

"Well, why would you think that anything happened in your building?"

"Maybe because of all the stuff we published about Osama bin Laden," she said.

"Martha, do you really think Osama bin Laden reads your tabloids?"

"I don't know, but we've been real hard on him, and you never know."

In the next couple of days, Martha Moffett's symptoms subsided, but she remained unsettled. She recalled:

> I kept watching the news. They were saying that Bob Stevens was an isolated case, that he probably caught it in the woods, and so forth. But friends were calling me and saying, "Why would you have pneumonia on this particular day? Are you sure you don't have anthrax?" I was pretty confident I had the right diagnosis or I would have had other symptoms. But then I was also aware that our mailman, Ernie Blanco, was in a hospital with pneumonia. It all became like, "What's going on? What's going on?"

Martha Moffett did not have anthrax. But on Sunday, the day she left the hospital, her suspicions that American Media may have been a target seemed validated. Laboratory tests had confirmed that anthrax was present in several locations in the AMI building. At about 6 p.m., Jean Malecki, director of the Palm Beach County Health Department, ordered the building closed. Neither Martha Moffett nor any of the other 500-odd AMI employees would be allowed back in, perhaps ever again.

That evening, working past midnight from the health department's emergency operations center in West Palm Beach, Daniel Rotstein phoned the company's managers. "I woke up a lot of people and asked them to call their staffs." He also asked them to identify recent visitors to the building, anyone who had been inside since August 1. They were stunned to hear him say: "You need to tell everyone to go to the health department offices in Delray Beach tomorrow morning for testing and antibiotics."

Martha Warwick, an *Enquirer* associate editor, was jarred out of her sleep by the 1:30 a.m. call. As she heard the news, she thought, "My God, we've just become one of our own headlines." Her heart was pounding. "I was in shock," she said. She thought of her three youngsters. "I frantically tried to remember when my children had been in the building, but I was having a hard time thinking clearly." Then she realized the date. They had been clamoring to see her office, and she had taken them over on the last day of their summer vacation, August 13.

Early Monday morning Warwick and her children joined the worried crowd that had begun to gather outside the Delray facility, just north of Boca Raton. In the course of the day, some 1,000 people who were considered at risk had their nostrils swabbed and were given a 10-day supply of antibiotics. Anxiety was rampant, even among the investigators. FBI agent Judy Orihuela, who worked in the bureau's Miami office, asked one of the health officials what it took to get anthrax. "When he said, 'Has to be 8,000 to 10,000 spores,' I was a little nervous." She knew how tiny that amount actually was. "You never know when you're going to come in contact with it."

When Martha Moffett left the JFK Medical Center, she departed through the hospital's main lobby. On her left she passed two elegant wing chairs on either side of an Early American mahogany chest. Above the chairs and chest hung pictures of tropical flora. Still higher, a prominent gold inscription engraved into the white wall said: The Generoso Pope Jr. Lobby.

Reflecting his first name, Mr. Pope had been the hospital's most generous benefactor. A contributor to causes ranging from the American Heart Association to the Chamber of Commerce, "Gene" delighted in favoring the JFK center. Beginning in the 1970s until his death in 1988, his cumulative gifts to the hospital approached $10 million. The JFK Medical Center was both his principal community cause and the hospital he himself used. Conveniently located, it was 15 minutes from his home in the posh community of Manalapan and 5 minutes from his office in Lantana. In the end, Pope died of a heart attack on the way from his home to his hospital in an ambulance he had donated.

Gene's father, Generoso Pope, Sr., had emigrated as a poor boy from Italy to the United States in 1906. He later became the wealthy owner of *Il Progresso*, the largest Italian-language newspaper in the United States. When he died in 1950, Gene, then 23 and a graduate of the Massachusetts Institute of Technology, became the publisher. But other family members gained control and pushed him aside. Suspicions about the senior Genoroso's ties to the Mafia trailed his son—for example, Frank Costello, the "John Gotti" of his time, was a family friend. In 1952, Costello lent Gene $25,000 to purchase a struggling weekly broadsheet whose main appeal was its racing tips. Over the next few years, circulation remained stagnant at 17,000. The paper survived only with more help from Mr. Costello and his associates. In turn, Gene obligingly published the winning numbers from the Italian lottery, around which the American Mafia had developed its own numbers racket.

One evening in the late 1950s, while driving home in New Jersey, traffic slowed to a crawl. Gene later realized that the long delay was due less to the accident up ahead than to rubbernecking. He was inspired. Blood, guts, gore—that's what would sell his paper. The weekly he had bought, the *New York Enquirer*, took on a new focus and before long had a new name: the *National Enquirer*.

The paper's rebirth began a 10-year period that journalist Jonathan Mahler calls the "gore era." Typical headlines read: "Violent Criminal Kills Pal and Eats Pieces of His Flesh," and "I Cut out Her Heart and Stomped on It." The new incarnation drew a national readership of 1 million. But by the end of the 1960s, circulation was flat, and Pope began searching for a formula with broader appeal. In the mid-1970s he found one: celebrities—their dirty laundry, their successes, their secrets, their joys, their embarrassments. And whatever the written words, nothing would beat the choice, sometimes lurid, photograph.

Soon after Pope's 1970s epiphany, he moved the *National Enquirer's* offices from New York to a one-story building in Lantana. Working 7 days a week, he nourished his new focus on celebrity foibles, and before long circulation mushroomed to 6 million. The driving aim was to get the scoop, and the *Enquirer* would do whatever was necessary to get it. In 1982, for example, $15,000 bought an exclusive from a gardener who had been within earshot of Princess Grace's last words after her fatal car accident in Monaco. For $60,000 the *Enquirer* obtained the photograph that derailed Gary Hart's 1988 presidential candidacy. After Hart denied he was having an extra-marital affair, the paper printed the famous shot of Donna Rice sitting on his lap aboard a yacht named "Monkey Business." Despite the paper's flamboyance and sensationalism and its payments to people for interviews, its stories essentially were built on facts.

In 1989, the year after Pope died, his family sold the *Enquirer* for $418 million. Circulation, which had begun to decline before his death, continued to fall. But several exclusives in the mid-1990s helped bring stability to the tabloid. It was the staff of the *Enquirer* who found pictures of O. J. Simpson wearing the Bruno Magli shoes he denied owning during his 1995 murder trial. It was on the cover of the *Enquire*r that viewers may have first seen the picture of President Clinton greeting Monica Lewinsky in a crowd. The paper had purchased the rights from *Time* (yes, *Time*) magazine to publish the picture. The picture also appeared on the cover of *Time*. The mainstream media were now covering subjects that previously had been left to the tabloids. But some tabloids had also changed. Stories in the *Enquirer*, if gossipy, were no longer just fanciful.

Loyalty among AMI employees is widespread. "We take our journalism very seriously," Ed Sigall has said. Now a senior editor, he joined the *Enquirer* staff in 1974 as an assistant editor. "We don't write about 'Elvis,' or things like that." Similarly, Martha Moffett, who retired in early 2002, believes that people who think poorly of the tabloid press are misguided. "It's interesting that within publishing the reputation is fine. The reputation is of a successful particular brand of publishing that is not outside the pale of the publishing industry."

Over the years other tabloids, such as the *Globe* and the *Star,* gained substantial readerships in their own right. But in 1999, David Pecker and Evercore Capital Partners brought them all under one management. For $835 million they acquired American Media, Inc., which already was publishing the *Enquirer* and the *Star* and now added Globe Communications, which owned the *Globe*, the *Sun*, and the *National Examiner*. The new owners also decided to move the company's headquarters from Lantana to the Globe building in Boca Raton. In January 2001, after a $12 million renovation, the spanking new insides were ready for occupancy. The refurbished three-story building, with 67,000 square feet of floor space, had become tabloid heaven. Displayed in the lobby were the current issues of virtually every tabloid weekly published in America. Ten months later, because of anthrax, the building was sealed shut; but by using other facilities, the papers were able to remain in print, and AMI could still proclaim that it "publishes seven of the top 15 weekly publications in North America."

According to Daniel Rotstein, each AMI weekly occupies a distinctive niche. But when he explains this, some don't sound very different from others: "The *Enquirer* is celebrity and human interest," he says. "The *Star* is mostly celebrity. The *Globe* is celebrity and some human interest but with a harder edge." In the *Sun* "you see more things about religion, horoscopes, prophesies, more unusual types of things."

Every AMI tabloid masthead begins with the same listing: David Pecker, chairman, president & chief executive officer. Next come the names of the paper's own staff. In the *Sun*, after Pecker's listing, that means Mike Irish, editor-in-chief. Farther down the masthead, in issues printed before AMI was visited with anthrax, were the names of the *Sun's* two photo assistant editors: Bob Stevens and Roz Suss.

Suss began working for the *Sun* in 1988, two years after she moved to Boca Raton from Miami. Although she had first arrived in Florida as a young bride in the 1960s, her Baltimore twang still rings clear. Now a graying grandmother, her small face turns mournful as she speaks of Bob Stevens. They had sat little more than arm's length from each other. "He was a kind, generous, funny man. Absolutely the sweetest." Nodding in agreement, Joan Berkley, the *Sun's* office manager, adds: "Everybody liked him. Everybody."

Suss and Berkley offered their observations in a room off to the side of the large, open space that AMI has been renting since the end of 2001. It was April 2002, and hundreds of employees were tapping out tabloid stories in a newsroom the size of a football field. Unlike in the three-story building that had to be abandoned in October, all the new offices are on the ground floor. No one is sure yet whether the company's own building will be decontaminated before the 2-year lease here ends. The current quarters, on Communications Avenue, are a half mile down the road from the AMI building on Broken Sound Boulevard. Though the working area is large, the floor space is less than half the 67,000 square feet of the old building.

For 2 months, before the rental arrangements for the new space were completed, the AMI tabloids were produced at scattered AMI facilities. Some were processed in the former *Enquirer* building in Lantana, others in offices in New York. The *Sun* and the *Weekly World News* were assigned space in the Miami building that produces *Mira!*, an AMI Spanish-language tabloid. A few days after the main building was closed, FBI agents visited the *Sun's* office in Miami. "They interviewed each of us individually for perhaps 30 or 45 minutes," said Roz Suss. She was interviewed by two agents and later received calls from other agents "to go over parts of what I said."

By late 2001 the FBI was consumed with the issue of terrorism. Special agent Judy Orihuela, who had spent most of her 11 years with the bureau working on bank fraud, recalled that soon after the anthrax diagnosis, "nobody was sure of anything." Now the media spokeswoman for the bureau's Miami division, she said, "At that point, in October, it was like 100 percent of our agents were doing either 9/11 or anthrax, one of the two." The division's 400 agents were reinforced by scores more sent from Washington and other offices. About a hundred agents alone were combing through the AMI building. Others were conducting interviews, and still others were planning and analyzing information back in the office.

Located in an industrial area near Interstate 95, the FBI's Miami building is ringed by dozens of 4-foot-tall cement planters. The bulky gray objects stand as sturdy barriers against any unwanted vehicle. Orihuela's third-floor office is adjacent to a long conference room, where in April 2002 she reviewed with me the events that took place 6 months earlier. Dressed in a red jacket and black pants, she glanced at the two dozen empty chairs around the oval table and the 20 others against the wall.

On Sunday, October 7, she was at home playing with her daughter when the phone rang. "You need to come down here," the voice from her office said. FBI agents had already been accompanying search teams looking for the cause of Bob Stevens's anthrax. When she arrived at the office she learned that spores had been found in the AMI building.

> It was weird. I remember we were in this conference room and everybody was here. All the top management, a lot of agents that were assigned to WMD [weapons of mass destruction], some terrorism people, agents from West Palm Beach.

Orihuela looked around the conference room as if to recapture that Sunday moment.

> I think a CDC person was here too, plus we had somebody on the speaker phone, a woman doctor from the CDC or the health department. We had attorneys from the U.S. Attorney General's office here, too. Because they had found that Stevens's desk was contaminated, we knew it was going to be a criminal investigation.

Roz Suss told her FBI interviewers about a strange letter that had arrived at the *Sun* weeks earlier. "I saw what I thought turned out to be the anthrax because of the way Bob Stevens handled this particular thing." A week or so after the initial interviews in Miami with the *Sun's* staff, the FBI summoned people from the paper to the bureau's office in West Palm Beach. Suss estimated that about 10 *Sun* staffers were there. "They asked each of us to write down what we saw, and they talked with us as a group." The accounts sometimes varied. "I think this helped convince them that there was more than one letter or package with anthrax."

Although holding the same job title as Bob Stevens, that of assistant photo editor, Suss acknowledged that she shared none of his artistic talent. Rather, her forte is her recall ability. "My eyes are my whole job and my memory." Suss's work includes selecting photographs that are appropriate to stories. Over the years the paper had accumulated files containing thousands of pictures. "They rely on me to remember what was in these files," she said, "and I could always tell them what file had what picture in it." Joan Berkley, the *Sun's* office manager, nodded in agreement and said, "I have to tell you, she has a fantastic memory." Suss's recollections, then, would seem to deserve particular respect.

"From my vantage—I'm down in the photo section with Bob Stevens—I remember Bobby Bender bringing a letter over." Bender, a news assistant, had been working for the *Sun* for 2 years. His job was to find stories for the paper from the Internet and other newspapers and to draw from leads that might come in the mail. His desk was at the far end of the room containing the *Sun's* two dozen cubicles and offices, about 40 feet from the photo section. "The letter was trifolded, and he had it in his hands this way," Suss said, as she drew her hands together, palms up. "I looked up and just watched the scenario occur with the letter in his hands."

"Look at this crazy letter that came to Jennifer Lopez," Bender said, according to Suss. He said that the letter had been sent to Lopez care of the *Sun*. The letter's message was apparently aimed at dissuading Lopez from a forthcoming marriage, but no one is sure of its text now. Suss did not see the words in the letter, but she did see that it contained powder.

Bob Stevens had just gotten up from his desk and was walking to the hallway when Bender rounded the corner into the photo section. Stevens, apparently heading for the library, according to Suss, stopped when he came upon Bender. "Let me see that," Stevens said. Suss describes Bender's handover to Stevens as "sort of like an automatic exchange. Bob took the letter from Bobby and held it in the same manner that Bobby was holding it—with whatever was in it." Bob Stevens walked back to his desk. He sat down with the letter over his keyboard, looking into it. Suss's chair was 6 feet away. Behind the L-shaped desks in their cubicle, Stevens faced north and Roz Suss east. "I'd seen all kinds of things drop out of letters and all kinds of junk, so I wasn't especially curious," she said, but then Stevens turned his head and looked over his shoulder at her:

> With that Bob says to me, "Hey, I think there's something gold in here. It looks like a Jewish star sticking out of the powder." I walked up behind him and reached over his shoulder. I pulled this little star out of what looked like a mound of powder in this letter. I remember it as a fine white powder.

Suss's fingers barely touched the powder as she plucked the charm out. "It looks like something from a Cracker Jacks box," she observed to Stevens. She examined it for a moment longer. It was a tiny plastic gold-colored Star of David with a loop attached. Stevens still seemed absorbed by the letter as she turned away. "I threw the star in my wastebasket and went back to the work I was doing."

Carla Chadick, a staff writer whose desk was nearby, also came over to see the letter. She remembers the powder and the charm. She saw the first line or two but does not recall whether the writing was in script or block lettering. "It was nothing that really struck me. It was just a cursory glance. We get these kind of letters all the time." She does remember thinking: "How stupid of them to send it to the *Sun*." And she chuckled to Bob Stevens: "We don't do celebrities. They should have sent it to the *Globe*." Roz Suss also recalled Chadick saying, "What crazy nut would send us something for Jennifer Lopez when we don't even deal with celebrities?"

Stevens joined in the mirth. "Boy, we get idiot letters, but this one was really off," he said. After a few more laughs, Chadick went back to her desk and resumed work. So did everyone else. "That was the end of it as far as I was concerned. It was just such a nonissue," she said. Stevens surely thought the same and probably tossed the letter into the trash, which was routinely incinerated.

There is no doubt that the incident took place on September 19. That is because on that day, Joan Berkley, who usually receives and opens mail for the *Sun*, was absent. Had she been at work, she would have been aware of the letter and its contents. In fact, as Roz Suss was looking over Stevens's shoulder, she recalled thinking at that moment that, if Joan Berkley had been here, she would have been the first to see the letter.

If the powder in that letter was anthrax, it was almost surely what killed Robert Stevens. Symptoms of inhalation anthrax are likely to occur during the first week or so after exposure, though they may appear up to 60 days later. Stevens's symptoms began around September 29, 10 days after he hovered over the powder.

Assuming the powder was anthrax, the fact that neither Roz Suss nor Carla Chadick became ill could mean that they did not inhale as much as Bob Stevens. It could mean that despite their exposures, their bodies were more resistant. Or it could mean that their incubation period was longer and that, had they not started to take antibiotics on October 8, they too would have come down with the disease.

Complicating their contentions, however, is the fact that Bobby Bender disputes much of Suss's account. After listening quietly to Suss's and Berkley's recollections, he said he does not remember giving a letter to Stevens. Rather, he recalls handling a large envelope or box addressed to Jennifer Lopez, care of the *Sun*. In it, he says, was a cigar tube containing a cigar, a small Star of David charm, and something that seemed like soap powder. "You could smell the powdered detergent," he said. In the middle of his description Joan Berkley turned to him: "Bobby, there were two different letters or boxes, because I remember that cigar. I remember that one. I was out for the other one." Suss then said: "See, this is why the FBI felt there was more than one package. I never saw this package or anything related to this package."

FBI agent Judy Orihuela also referred to Bender's presumably handling "one of the Lopez letters." The likelihood of more than one mail item is strengthened by Jean Malecki, the county health department director. The epidemiological investigation pointed to "at least two anthrax letters sent to AMI," she said. The investigation turned up:

> . . . two different routes that would eventually go back to the AMI building. AMI is a conglomerate. There was the *National Enquirer*, which had its own major building in the Lantana area for years. There is still mail that goes to that old building and eventually comes back down a pathway that goes down to AMI. The pathway to the *National Enquirer* was contaminated with powder. There were two separately contaminated facilities. That's been independently corroborated by interviewing people.

The fact that an anthrax letter went to the old *Enquirer* building in Lantana raises provocative questions. Beginning in January 2001, the address listed for all AMI publications was that of the refurbished AMI headquarters in Boca Raton—previously the Globe Communications building. Not since December 2000 had the masthead of the *National Enquirer* listed the Lantana address. Whoever targeted the *Enquirer* apparently would have taken the address from a very old copy. Or perhaps not. If the mailer lived in the Boca/Delray/West Palm area in 2001, he would have had access to the *Bell South* telephone directory for those communities. In the business section of the 2001-2002 directory, the address for the *National Enquirer* is listed as 600 S.E. Coast Avenue, Lantana 33462. In contrast, the address for the *Sun* was that of the building in Boca Raton as it always had been, because the *Sun* had been owned by the Globe company.

Thus, if the anthrax mailer obtained the *Enquirer's* address from current issues or from the Internet, the letter would have gone to the Broken Sound Boulevard address in Boca Raton. The only current source that still lists the old Lantana address is the telephone directory. And the directory is easily accessible only to people who live or work in the area.

In 1974, when Bob Stevens began to work for the *National Enquirer*, among the first people he met there was Mike Irish. Three years earlier, Irish, like Stevens, had emigrated from England, and became assistant managing editor. The connection between the two men led to Mike's wife, Gloria, a real estate agent, finding a home for Bob and Maureen. The house she found for them was the one in Lantana, where Bob lived until his death.

Through the years Mike Irish held various positions with different tabloids before becoming the *Sun's* chief editor. His round face accentuates his cherubic appearance. The most visible spot of hair in the vicinity of his totally bald pate is his faint gray mustache, which curls upward with his smile. He finds no pleasure in recalling the particular problems he and his wife faced after September 11.

The Irishes live in Delray Beach, 15 minutes from the AMI building in Boca Raton. Behind the wooden fence in front of their house, the pathway is surrounded by a lush garden of tropical plants and flowers. A large piano and soft sofa are centerpieces of the living room. Gloria spoke softly. She told of the trauma she shared with all Americans as she learned about the hijacked planes crashing into the World Trade Center. At 10:15 that September 11 night, Mike and Gloria were in bed watching the television news when the phone rang. It was the FBI.

"Mrs. Irish, did you rent an apartment to Hamza Alghamdi?"

"Yes, I did." Gloria was not entirely surprised by the question. When she heard earlier in the day that some of the suspected Arab hijackers had been living in South Florida, she thought about the two men to whom she had shown apartments. "And I also rented another apartment to his friend, Marwan Al-Shehhi. Marwan told me he was a pilot and that he was taking flying lessons."

"Well, we'd really like to talk to you now."

"Now?"

"Yes. Can we meet you at your office?"

Gloria and Mike dressed, drove a few blocks to Federal Highway, and turned left. Three minutes later they were in front of Pelican Properties, Gloria's real estate company on South East 6th Avenue. She unlocked the door, turned on the lights, and waited for the agents.

The FBI had identified several of the apparent hijackers from their airplane tickets. The bureau was now looking for people who might have been in contact with them. Gloria's name came up because records in the local homeowners association indicated she had been the rental agent for Alghamdi.

"How did you come to meet them?" one of the two agents asked. "They just walked in," Gloria replied. "It must have been in May." The men wanted to rent apartments near each other for June and July. During the next couple of weeks Gloria showed them dozens of places before they found any to their liking. The apartments they settled on were in Delray Beach, Alghamdi's at the Delray Racquet Club and Al-Shehhi's at the Hamlet, an attractive golf course community. The apartments each rented for about $1,000 a month. The men paid the full amount in cash plus a security deposit of $1,000 for each apartment. Later, Al-Shehhi's lease was extended to August 12, Alghamdi's to August 30.

"What were they like?" the agents asked.

"Well, Marwan Al-Shehhi always came in with a smile. If they were going to be 2 minutes late, he'd call. He was very considerate," she said.

"And Hamza Alghamdi?"

"He was weird," Gloria answered. "Hamza just stared and never spoke to me. Marwan said it was because he didn't speak English." Alghamdi's fixed gaze made Gloria uncomfortable. Later, it dawned on her that Marwan never translated anything for him and that he must have understood what was being said.

"Did they indicate why they wanted to live here?" Al-Shehhi told Gloria they had been living elsewhere in Florida—on the west coast and later south in Coral Gables. He was now taking flying lessons in the Delray area and wanted to live nearby. "He said he was doing hours so he could get his commercial license," Gloria recalled. When she heard about the flying, she told Al-Shehhi that her husband, Mike, held a private pilot's license. She had many conversations with Al-Shehhi, although mostly about his eagerness to find the apartments. "I mean, Marwan called me all the time."

After the FBI agents finished questioning her, they asked for copies of everything in her files related to the two rentals. She pulled out the lease agreements, payment records, and copies of the two men's driver's licenses. She found a copy of a blank check, which she recopied for the agents. She first saw the check when Al-Shehhi had asked how the security deposit would be refunded after they vacated the apartments. Gloria said, "If you give me a deposit ticket for your bank account, when I get your refund, I'll deposit it to your account for you." Al-Shehhi replied, "I don't have a deposit ticket with me, but I have a check." She said, "That's fine, I'll make a copy of it so I'll have your account number."

Sometime after the visit from the FBI, after the names of the hijackers had been made public, she looked again at the blank check. The account was with the SunTrust Bank, Gulf Coast Downtown Venice Office, and the owner's address was 4890 Pompano Road, Venice, Florida. The name that followed Al-Shehhi's on the check had previously meant nothing to her. Now, upon seeing it, she gasped. It was Mohammed Atta, the alleged leader of the September 11 attacks who piloted one of the planes into the World Trade Center. Marwan Al-Shehhi piloted the other. Atta was also among a group of Arab men who, the previous month, approached airplane mechanics in Florida to ask about cropduster specifications, including their carrying capacity. Cropdusters are considered potential weapons if used as vehicles to spread biological or chemical agents over large areas. Upon learning on September 23 about the men's inquiry, the FBI ordered the temporary grounding of all cropdusters in the United States.

As distressing as September 11 was for Gloria and Mike Irish, the anthrax incidents in October turned their lives upside down. "It was awful. It was unbelievable," said Mike Irish. Bob Stevens had died of anthrax from a letter apparently addressed to the *Sun*. Mike was editor-in-chief of the *Sun*. According to the FBI, as many as 15 of the 19 hijackers may have lived in Florida. Six of them had addresses in Delray Beach or Fort Lauderdale, a few miles from the AMI building where the *Sun* was published.

Mike's wife was the real estate broker who rented to two of the hijackers. Was there a relationship between the airline terror and the anthrax terror that was somehow connected to the Irishes or to AMI? The media were certainly looking for one, Mike said. "In the course of a week after the anthrax attack, she and I handled 60 or 70 calls. We steadfastly were saying, 'We have nothing to say. Goodbye.'" When a reporter wrote a story falsely claiming that he had an exclusive interview with Gloria, she was deeply hurt. "It crushed her. She couldn't sleep," Mike recalled. He also recognized the irony in his annoyance: "I have to say, since I'm part of the media, that I can't blame them. I kind of hate them for it, but I can't blame them for trying to build a story, but in fact there was none."

Could Gloria Irish possibly have mentioned to her two clients that her husband worked for American Media or for the *Sun*? Doubtful, she said, because the two men were focused on looking for apartments. "They couldn't care less about anything else." Gloria acknowledged that she commonly talked about her husband with other clients, but she cannot recall doing so with the two Middle Eastern men. Yet at another point in her conversation she did concede telling them that Mike was a pilot.

Did Al-Shehhi know where she lived? "Normally, I never have any problems in giving my home phone number to a client, but I don't think there would have been any reason I gave it to them." So she did not give them her home phone number? "There is the possibility I did, but I don't think so." Gloria remembers that after they signed up for the Delray apartments, "I almost told them where I live, but I didn't. I did tell them I live a couple of blocks from them, and if you need anything, tell me."

Mike Irish is adamant that all the apparent dots are unconnected. "It was just a total coincidence," he said. He may be correct. Still, as long as the identity of the perpetrators remains uncertain, perhaps Martha Moffett's suspicions should not be dismissed. The AMI librarian was among the first to suspect a connection between the bin Laden terrorists and the anthrax terrorists.

Here's another dot: the Delray-area phone directory carried only one listing of an "Irish" on Pelican Way, and the first initial is "M." Gloria had told Al-Shehhi and Alghamdi that she lived and worked (at Pelican Properties) not far from their apartments. One look at the phone book would have pinpointed her home address, phone number, and husband's first initial. What significance could an "M. Irish" have for anyone not familiar with the *Sun*? Probably none. But for someone who regularly read the newspaper, the name "Mike Irish, Editor-in-Chief," could be seen every week. It was at the top of the masthead, just below David Pecker's name. And for anyone interested, the masthead also displayed the paper's address in Boca Raton.

On October 10, 2001, an article appeared in the *Miami Herald* with the headline, "Authorities Trace Anthrax that Killed Florida

Man to Iowa Lab." It contained an intriguing piece of news: "Investigators confirmed that two hijackers who died in the Sept. 11 terrorist attacks had subscriptions to tabloid newspapers published in the Boca Raton headquarters of American Media, Inc., where photo editor Robert Stevens is believed to have contracted the fatal disease." Nine months later, when asked about this report, the FBI would neither confirm nor deny that the hijackers had subscriptions to the tabloids. Nor would an AMI spokesman comment on the matter because, he said, the case was still under criminal investigation.

The *Sun*, with a national circulation of 226,000, is among the smallest of the AMI papers. Its Florida circulation is about 15,000, and almost all are single-copy sales. In 1999 subscriptions throughout the state numbered only 253. If any hijackers were among the subscribers, this would advance the notion of their connection to the anthrax perpetrators. It suggests a plausible explanation of why this relatively obscure tabloid might have been the target of an anthrax letter.

During the week after Bob Stevens died, the nasal swab tests conducted on the AMI employees found anthrax spores on one of them—Stephanie Dailey, 36, who had worked in the mailroom. By the time her exposure was confirmed, she, like the rest of the AMI staff, had already been on Cipro for 2 days. She never developed symptoms of the disease. Another bit of encouraging news was that by the end of that week Ernie Blanco began to improve. But the good news about Dailey and Blanco was quickly overshadowed by news of more anthrax horrors elsewhere in the country.

CHAPTER THREE

THE NATION AT RISK

At 11:40 a.m. on October 10, 2001, work came to a halt at New Jersey's Election Law Enforcement Commission in Trenton. A staff member had just opened a letter that contained a white powdery substance. A call to the state police brought a firm instruction: "Don't anybody leave." Within minutes more than 50 police, fire, emergency medical, and environmental staff began to arrive. The block in front of the building was sealed with yellow crime tape. Located on the thirteenth floor at 28 West State Street, three blocks from the state capitol, the commission administers New Jersey's tangle of election laws and procedures. The 300 workers inside the 14-story structure were told to remain in their offices, to close their windows, and to not eat or drink anything.

Three hours later, after preliminary testing of the powder, state police spokesman John Hagerty announced: "It is not anthrax." He did not yet know what the material was, only that it was "not chemicals that are life threatening. We have a hoax situation." That week, similar scare scenarios were taking place elsewhere around the country. Disruptive threats, all false, had occurred in Burlington, Vermont, in Covington, Kentucky, in Darien, Connecticut, and as far away as Honolulu, Hawaii. The most jittery reactions, not surprisingly, were in South Florida, where anthrax had actually been found. "It gets me a little worried—spores flying in the air and all that," said Steve Siebert, a carpenter who worked near the American Media building in Boca Raton. "My only fear is that I've got it and don't know it." Jan Schuman, manager of the Boca Pharmacy said: "It's Cipro, Cipro, Cipro," as he described the surge in demand for the primary antibiotic used to treat anthrax. His store, which usually dispensed 200 tablets a week, had just put in an order for 600. "It's out of control. People are frantic."

Acute frenzy was largely limited to South Florida. The rest of the country remained wary but hopeful that actual exposures to the bacteria had been limited to the Florida area. One week after Bob Stevens's death, that hope proved to be in vain.

At 3 a.m. on Friday, October 12, a telephone call jolted Marcelle Layton out of her sleep. David Ashford was on the line from the Centers for Disease Control and Prevention in Atlanta. "Marci, I'm calling to give you a chance to wake up. I'll be phoning back in 5 minutes. Jeff Koplan wants to talk to you and some others on a conference call." Dr. Layton, New York City's assistant health commissioner for communicable diseases, had gone to bed an hour earlier "with a smile on my face," she recalled. She had left her office after midnight following a good-news phone call from the CDC. The PCR (polymerase chain reaction) results for a skin biopsy of an NBC employee were negative for anthrax. Happily, the CDC had confirmed the city's own negative findings. But the Atlanta-based agency was still performing capsular and cell wall testing on the specimen—the same kind of tests Phil Lee had done in Florida a week earlier on Bob Stevens's sample. Still, Layton felt optimistic.

Then came Ashford's ominous middle-of-the-night call. When the phone rang again, Jeffrey Koplan, CDC's director, came on to say that the immunohistochemical staining for the specimen was positive. (Capsular and cell wall tests both involve immunological and chemical staining, thus the shorthand term "immunohistochemical" testing.)

When Phil Lee had conducted these tests on Stevens's sample, he had come away from his microscope knowing he had seen something momentous. Now, in Atlanta, Koplan and two other CDC officials—Sherif Zaki, the lead pathologist for the agency, and James Hughes, director of its infectious diseases center—looked through the microscope and experienced the same sense of gravity. Zaki, who had devised these anthrax-specific tests a few years earlier, told Koplan he was 90 percent certain that the results were positive. The anthrax outbreak, evidently, had not been confined to Florida. Still, Koplan seemed reluctant to make a definitive pronouncement. Later that morning he was on the phone with New York's Mayor Rudy Giuliani, who asked, "Are you sure it's anthrax?" "Well, we have a high degree of probability," Koplan replied. Hughes recalls that moment vividly. A year later he told me: "We were reluctant to say 'absolutely' because there is always the chance of error. If we say 'yes,' and it turns out not to be anthrax, the consequences would be huge."

"No, no, no, don't give me that stuff," the mayor responded to Koplan. Giuliani, whose decisive leadership in the wake of September 11 had become legendary, would not settle for probability. "Is it anthrax or is it not?" he insisted.

"Yes," Dr. Koplan said.

"Fine, that's all I need to hear."

Months later Dr. Layton reflected on that day and the weeks leading to it. Her third-floor office on Worth Street in lower Manhattan is nine blocks north of the abyss where the Twin Towers had stood. Behind her desk are shelves of volumes that befit her specialty in epidemiology—texts on infectious diseases, vaccines, AIDS, a folder of CDC reports. The sound of classical music, the single feature in her office not directly related to her professional charge, is faintly audible. "I never met anyone who works as hard," Tim Holtz, a staff physician, observed when Layton was out of earshot. "She is always here."

Layton evinced a shy smile as she mentioned that her responsibilities include "disaster planning" for bioterrorism. But on September 11 everyone in the department was consumed with overseeing the city's health response to the morning's horror. The office phones were down, so communications went out erratically by cell phone, computer, and hand delivery. "I'm the chair of the surveillance committee," Layton said. That day, from her small corner office, she directed the citywide surveillance of hospital emergency room activities, injuries, staff responses, and availability of equipment. By all accounts, the public health effort was remarkably successful. Doctors and hospital staff cared for the surge of patients with few hitches. Layton retrospectively sorted out her feelings:

> When you're in the midst of an emergency, I mean you're so focused. I didn't see the physical destruction. We couldn't even watch it on TV because our reception went down. I saw the hole in the first tower and that's the last thing I remember seeing. My memory is just full of

> a nonstop effort to get things organized in an extremely chaotic situation and not fully comprehending what was going on.

During that day, Layton and Joel Ackelsberg, the city's deputy commissioner responsible for coordinating a bioterrorism response, reminded each other that at some point they should talk about the threat of a germ attack. The first chance came in the early hours of the next morning. "At 2 a.m., four or five of us, including a couple of people from the CDC, met here in my office," Layton said. "We started talking about what we needed to do to be on the alert for any bioterrorist event." In cooperation with the CDC, they planned for heightened surveillance at the 29 hospital emergency departments in New York City and for getting the word to "our medical and laboratory communities to report any unusual cluster or disease manifestations."

Then, 4 weeks later on October 12, came the 3 a.m. phone call from Jeffrey Koplan. The fact that someone in New York had contracted skin anthrax, apparently a victim of bioterrorism, further shocked the country. In truth, the city was better prepared to cope than many other communities. Local police, firefighters and medical people had all participated in mock attacks. Still, confirmation of a skin anthrax case came as a surprise. "All our efforts in the past several years had been focused on the expected covert release of an aerosolized agent, so we were focused on an inhalational outbreak," Layton said. The manner of delivery was also a surprise. No one expected that the threat letters were a likely way to cause disease.

Later that day the city issued a statement under the heading: "Health Department Announces Anthrax Case in New York City." The text explained that an NBC employee had contracted cutaneous anthrax. Although labeled an "alert," the statement seemed almost reassuring: "The employee has been taking antibiotics since October 1 and is doing well."

But the Florida anthrax incidents and the attack on the World Trade Center had just occurred and New Yorkers were shaken. Mayor Giuliani echoed the U.S. Postal Service's warning that people should not open suspicious packages, such as those without return addresses. The NBC studios had been sealed, and internal mail delivery was halted at CBS, ABC, CNN, and the Associated Press. Judith Miller, a *New York Times* reporter, had received a threat letter containing powder. Although the powder proved to be harmless, it intensified the sense of siege. From the *Washington Post*:

> And at the *New York Times*, where police sealed the building for several hours, employees stuck outside broke into tears and phoned family. Rhonda Cole, a senior sales executive at the newspaper, returned from a meeting to find the building locked down. For the first time, the cumulative events of the past weeks took a toll. "I'm shaky. I'm weak," said Cole, as she stood behind a police barrier on Broadway and watched hazardous-materials crews walk into her building. "Now I'm afraid. It's where I work. After a while, it's too much."

The confirmed anthrax case was that of Erin O'Connor, a 38-year-old assistant to Tom Brokaw at NBC. She remembered first seeing a sore on her chest on September 25. By the time she visited her doctor on October 1, the oval-shaped lesion, about an inch long, had become ulcerative. The lymph nodes in her neck were swollen, and she was feeling weak and tired. Dr. Richard Fried, her Manhattan internist, initially thought she might have an infection from a spider bite. When O'Connor told him about having opened a threat letter containing powder, he considered anthrax a possibility. But he did not share his suspicion with her because "I just did not want to alarm her." Nevertheless, he prescribed Cipro as a precaution, called the city health department, and sent a skin sample to the department's laboratory.

Joel Ackelsberg, who took Fried's call at the health department, doubted that the lesion was anthrax. When he heard about a suspicious letter, he thought about the scores of hoax letters the department had seen until that time. "All these events had been hoaxes, and my approach was that they were going to continue to be hoaxes." Dr. Ackelsberg recounted his feelings from his compact office on the second floor of the health department building on Worth Street: "The lesion was on the left side of her chest below her shoulder. It was in an unlikely location. We thought it was a spider bite that was infected. So I said, 'Let's just test the letter.'"

The health department contacted the Federal Bureau of Investigation, which managed to retrieve the suspected letter from a batch of hate mail that NBC had collected. The letter was postmarked September 25 from St. Petersburg, Florida, and said: "The unthinkable. See what happens next." But the health department lab was unable to grow anthrax from either the biopsy or the letter, and, said Marci Layton, "We sort of moved on to other things."

O'Connor's lesion later turned into a coal-black crust. (Anthrax derives its name from the Greek word for coal, *anthracis*.) When O'Connor heard about the anthrax case in Florida, she searched the Internet, wondering if her own lesion might be anthrax. On October 9 she visited Dr. Marc Grossman, a dermatologist at Columbia-Presbyterian Medical Center, who also thought it was anthrax. Her skin sample and the letter in question were sent to the CDC for testing. Then came the call from Koplan to Marci Layton on October 12. The letter did not contain anthrax, but the biopsy did. Health department staff prepared to go to NBC to take nasal swabs and other samples. Layton explained:

> We mobilized early that morning, and by noon we were on site beginning our investigation at NBC. That included looking for [the source] since we knew the initial letter was negative. We found a second letter thanks to one of the interns that worked with [Erin O'Connor]. It was found, brought to our lab, and tested positive for anthrax.

Postmarked September 18 from Trenton, New Jersey, the newly identified letter and envelope contained hand-printed capital lettering. The envelope, which contained no return address, was addressed to:

TOM BROKAW
NBC TV
30 ROCKEFELLER PLAZA
NEW YORK NY 10112.

O'Connor did not remember seeing this letter before, but she must have come in contact with it between September 19 and 25. The message on the sheet inside was a copy of the original, which evidently had been kept by the mailer. Under "09-11-01," the letter contained five lines, including the misspelling of "penicillin":

THIS IS NEXT
TAKE PENACILIN NOW
DEATH TO AMERICA
DEATH TO ISRAEL
ALLAH IS GREAT

In consultation with the CDC, the health department performed nasal swabs on more than 400 NBC employees. The "Nightly News" studio was closed. People who had visited the NBC offices between September 19 and 25 were urged to be tested and to begin taking an antibiotic.

New Yorkers were in for more bad news. On Monday, October 15, the health department announced the second case of cutaneous anthrax. Then on Thursday a third case and on Friday a fourth. The second case was especially heartbreaking. At 2 p.m. on Friday, September 28, a producer at ABC's World News Tonight had taken her 7-month-old baby for a visit to the company studio. By the time a babysitter took him home 2 hours later, several colleagues had cuddled and held him.

The next day, a red sore the size of a half-dollar appeared on the back of the infant's left arm. Thinking the lesion was caused by a spider bite, a doctor prescribed Benadryl, an antihistamine. The child's condition worsened, and on October 1 he was admitted to the New York University Medical Center. After 6 days of antibiotics, his arm started to heal, but his red blood cell count dropped and his kidneys began to fail. "He was in deep trouble," his mother said. With continued antibiotics and blood transfusions, after 6 days, the child's condition began to improve. A week later he was released from the hospital. Meanwhile, as stories of anthrax filled the airwaves, the child's doctor realized that his symptoms seemed consistent with the disease. The doctor notified the health department, and on October 13 a skin biopsy was sent to the CDC. Five days later the diagnosis was confirmed.

The third case was that of Claire Fletcher, 27, an assistant to Dan Rather at CBS News. She had developed facial lesions and a headache on October 1, was placed on antibiotics, and was doing well by the time of the diagnosis.

The fourth case, Johanna Huden, 30, was a *New York Post* employee who began to display symptoms on September 22, earlier than any of the other victims. Her finger lesion was treated with antibiotics, which started her recovery. Her belated diagnosis was confirmed on October 18 by tests of a skin sample. She had somehow been infected by spores from an unopened threat letter that had been stashed in a bin with other suspicious mail. The letter was later retrieved and found to test positive for anthrax. Besides the letter to Tom Brokaw, it was the only other piece of mail in New York that was identified as an anthrax letter. Like the Brokaw letter, it was postmarked September 18 from Trenton and was written in block capitals:

EDITOR
NEW YORK POST
1211 AVE. OF THE AMERICAS
NEW YORK NY 10036

The letter, which was a copy of an original, was identical to the one sent to Brokaw.

Amid the outbreak in New York City, on October 15 the Florida Department of Health announced that anthrax had been found in the Boca Raton post office. Spores were located in an area where mail was sorted for pickup by American Media and nearby buildings. This finding was another indication that the anthrax that infected the American Media building had been delivered by mail.

Peoples' worries about their own safety drew a guarded comment from Mayor Giuliani: "A balance has to be struck here between sufficient precautions and making people so frightened and so upset that they're not going to be able to conduct their lives, which means having people walking around in spacesuits all over New York." By the time the wave of cutaneous anthrax infections ended, seven people in New York City had been diagnosed with the disease. All recovered.

By October 25 nasal swabs had been taken on 2,580 people in New York City, and preventive antibiotics had been given to 1,306. All seven cases were connected to the media—two at NBC, three at the *New York Post*, one each at ABC and CBS. A letter to Tom Brokaw and one to the *Post* with the same postmark—September 18, Trenton, NJ—were recovered and found to have contained anthrax. While the cause of infections in the other New York offices was never found, the supposition is that mail laced with spores had been sent to them as well.

The anxiety generated by the anthrax cases in New York was intensified by news of anthrax outbreaks elsewhere. Erin O'Connor's skin anthrax, the first to be identified in New York City, was confirmed on October 12. That was just 8 days after anthrax was first identified in Florida. Then, 3 days later, spores were found in Washington, D.C.

At 7:30 a.m. on Monday, October 15, Dr. John Eisold climbed the steps of the U.S. Capitol's south entrance. As he had done every morning since he began working there in 1994, he headed down the corridor to his office. Along the way Eisold passed several statues—on his left, one of Harrison Schmidt, a former astronaut and senator from New Mexico; on his right, Philo Farnsworth, the inventor of television, and Father Damian, who ministered to lepers

in Africa. (They are among 100 statues in the building, two chosen by each state.) Two hundred feet from the entrance, Eisold turned left into a quiet alcove. A sign on the wall identifies a cluster of offices as the medical clinic. It is at this location that Dr. Eisold, the Capitol's chief physician, oversees a staff of three doctors, 15 nurses, and 15 corpsmen. Their potential clientele numbers 35,000, including the members of congress, employees, and visitors who might be on Capitol Hill on any day. Eisold and his staff routinely handle heart attacks, seizures, colds, fractures, sprains, abrasions.

He was at his desk at 10 a.m. when the emergency alarm sounded in the clinic. Eisold reached to his left for the phone. The call was from Senator Tom Daschle's office in the Hart Senate Office Building: "We have a letter here claiming to contain anthrax. There's powder in it." Six months after the call, Dr. Eisold leaned back in his chair and reviewed the unfolding events. He glanced toward a decorous fireplace beneath a large mirror on the far wall, emblematic of the building's 19th-century elegance. On a table behind him, just a swivel away, is his computer. "Since this was a biohazard event," he said, "we used a standard team of five people—a doctor and four others."

Dr. Norman Lee, whose office was across the hall from Eisold's, was dispatched as team leader. It was a drill he and the rest of the staff had rehearsed many times. Lee, like the other physicians in the clinic, is a naval officer, a lieutenant commander. Unlike Eisold, whose take-charge demeanor seems as much a reflection of personality as rank— Eisold is a vice-admiral—Lee is soft-spoken, diffident. "Before we left the office, when they said it was anthrax, I grabbed a big bottle of Cipro. We also brought doxycycline and enough culturettes for the swabs," said Lee. In two minutes an ambulance delivered the team five blocks away to the Hart Building on Constitution Avenue.

An intern had opened the threat letter in the sixth-floor mailroom of the senator's suite—the fifth and sixth floors are connected by an internal staircase. Like the letters to NBC and the *New York Post*, this one was also postmarked from Trenton, though stamped October 9, not September 18. In block capitals, the envelope was addressed to:

SENATOR DASCHLE
509 HART SENATE OFFICE BUILDING
WASHINGTON D.C. 20510-7020

Unlike the other two, it contained a return address:

4TH GRADE
GREENDALE SCHOOL
FRANKLIN PARK NJ 08852

Under the date, "09-11-01," the letter contained seven lines:

YOU CANNOT STOP US.
WE HAVE THIS ANTHRAX.
YOU DIE NOW.
ARE YOU AFRAID?
DEATH TO AMERICA.
DEATH TO ISRAEL.
ALLAH IS GREAT.

"When we got to his office, I was the only one from the team to go in the room because I'd already been vaccinated against anthrax," said Lee. The Capitol Hill Police's hazardous materials squad was already there, some in masks and protective outerwear. Their preliminary test showed that the powder could be anthrax. When Lee arrived, 13 of Daschle's staff members were in the room where the letter had been opened. The senator himself was at a meeting elsewhere. Lee ordered everyone out of the room and into the hallway. "I proceeded to test them with the nasal swabs. We told them to wash their hands and take their first dose of Cipro. They were clearly scared."

After the room was emptied, Lee turned to a masked HAZMAT worker. "I saw him standing over the bag containing the anthrax material that he had just wrapped. He showed me it was a small amount of powder, about this much." Lee curled his pointer finger into his thumb to make a small circle. "About the size of a grape," he said. The staff people began to pepper Dr. Lee with questions. "Do we need to go to the hospital?"

"Not unless you're having symptoms," Lee answered.

"What should we look out for?"

"Are you having any fever, any concurrent illnesses?" Lee responded to the group. No one indicated feeling sick, so he said: "Look, everybody's feeling fine. There's nothing to worry about. Take the antibiotic immediately. Your test results should come back within 48 hours or so."

A few people from Senator Russell Feingold's office, which adjoins Daschle's, may have been in their common hallway when the letter was discovered. The sixth-floor mailrooms of the two offices were separated by only a single wall, and the HAZMAT people realized that the staffs of both senators might need attention. They were all herded up to the ninth floor of the Hart Building.

There, in a large atrium, about 70 people stood waiting—members of Daschle's and Feingold's staffs, and a few others who had been in either senator's office. Dr. Lee and his team continued swabbing and distributing antibiotics, while police field tested samples of clothing. The anxiety and the questions continued. "Do I need a blood test? Should I tell my private physician?" Lee responded that, again, no blood test was called for in the absence of symptoms. There was no need at this time to call a private physician. He warned, however, that "we need to see every single one of you tomorrow to follow up." Each person had received only a couple of pills, and Lee reminded them, "We need to give you more antibiotics." Still, several people said they wanted to go to the hospital. Some were crying. The young woman intern who had opened the letter seemed especially distressed. Lee tried to help:

> I spoke to her. She was clearly scared. So they changed her clothes. The police took her clothing to sample it. Later, I spoke to her father also, to reassure him. Her father seemed okay, but I think he was somewhat shocked. He just said, "Should I be concerned?" I said, "Yes." I told him the test probably will come back positive. But I explained to him, "Even if the test comes back positive, it just means she's got some spores in her nasal passages. It does not imply disease nor future consequences." So I said, "Right now she does not need to go to the hospital. She needs to be at home with somebody."

The letter was brought to the U.S. Army Medical Research Institute of Infectious Diseases in Maryland where microbiologist John Ezzell later confirmed that the powder was indeed anthrax.

Meanwhile, Senator Feingold, who had been in Wisconsin for the weekend, was en route by car to Chicago's O'Hare Airport when his cell phone rang. Mary Murphy, his chief of staff, was calling from Washington with news about the anthrax. "They're quarantining our people with Senator Daschle's people up on the ninth floor of the Hart Building," she said. Feingold was shocked. He had just seen a television report that Tom Brokaw's office had received an anthrax letter. Now this. He felt as though he were under a dark cloud: "To be honest, I really felt then more afraid and concerned than I had at any point. I thought, 'Oh my God, this is really expanding.' I really felt concerned and sort of depressed about what was going on and worried about everybody." When Feingold landed at Washington National Airport that afternoon, he went straight to the U.S. Capitol where he had been scheduled to preside at a Senate session. He would not be able to return to his office in the Hart Building for 3 months.

A senior member of Senator Feingold's staff, who requested anonymity, reflected on that day. When she was notified about the suspicious letter her reaction was, "Jeez, I've got to deal with this. I hope nobody is going to be hurt or sick." Months later, she expressed "incredible pride" in the professional way the staff handled the situation. But she also witnessed unsettling glitches. She was among the people isolated on the ninth floor for swabbing, antibiotics, and a briefing. Expressing incredulity, she said, "But that afternoon, we were back in the office." Not until the next day, Tuesday, was the Hart Building ordered closed. What did Feingold's staff do that Monday afternoon? The tension was heavy, "so we tried to get people to relax." Feingold's old campaign ads were notoriously funny. "We crowded into the senator's office—he was away. We replayed his ads and kept everyone laughing."

The aide winced as she recalled that the door to Daschle's office "kept opening and closing all afternoon with people going in and out." The following week Senator Daschle announced that more than 6,000 people had undergone nasal swabs and that 28 had tested positive for anthrax. A breakdown of the swab results provided by the CDC indicated that 20 members of Daschle's staff had been exposed, two members of Feingold's staff, and six "responders" from the Capitol Hill Police. Some staffers accepted the offer to take anthrax vaccinations and others refused, but all remained on antibiotics for 90 days or longer. None became infected.

Meanwhile, Capitol Hill had become unnerved. A day after the discovery of the letter preliminary tests showed that spores had been found in the Senate mailroom. Attorney General John Ashcroft announced that the strain of anthrax found was "virulent, strong, very serious." The Capitol was shut down, and all six House and Senate office buildings were closed for screening. In subsequent weeks, suspicions of anthrax, many of them false alarms, prompted closings throughout the Washington area, including parts of the Federal Reserve Building, the State Department, and the Pentagon. A letter containing white powder prompted evacuation of the Supreme Court.

According to Feingold's aide, the senator's staff remained "anxious and edgy" for months. Only after the Hart Building was decontaminated and people returned in January 2002 did they seem more relaxed about the ordeal.

In 1986 a new mail-processing center opened in Washington, D.C., on Brentwood Road, 2 miles northeast of the Capitol off Rhode Island Avenue. Mail delivered to the Washington area, including Capitol Hill, was sorted at the center. During the week that the Daschle anthrax letter was discovered, four of Brentwood's 2,000 employees became linked by a common fate. The four men were black, as were most Brentwood workers, and were of similar ages: Leroy Richmond was 56 and Qieth McQue was 53. Thomas Morris was 55 and Joseph Curseen 47. All were bright and articulate: McQue was a graduate of the University of Maryland, Curseen of Marquette University. Susan Richmond, Leroy's wife, herself a postal worker at Brentwood, spoke of them all as "well-mannered guys." Jim Harper, who had operated a mail-sorting machine with Curseen, says that his friend Joe "would do anything for anybody at any time." He described all four as kind, good men.

They knew each other. "Rich" and "Mo," as Richmond and Morris were called, had been pinochle partners. During the 4 a.m. break in the night shift, Curseen led a Bible-study group. A half dozen men attended. "We'd read scripture and psalms and discuss what they meant to us," said Rich. Rich worked under McQue, who supervised the crew that picked up and sorted express mail from Baltimore-Washington International Airport. But the ultimate commonality of the four men began to surface on Tuesday, October 16, the day after the Daschle letter was opened. On that day each had begun to feel mildly ill.

Leroy Richmond, nearly 6 feet tall, is bone thin. His narrow face breaks into an easy smile even as he ponders that stressful week. By Friday he was feeling so achy that he did something he rarely does—"because I'm never sick." He called a doctor. That afternoon he went to the office of Dr. Michael Nguyen, an internist with Kaiser Permanente in Woodbridge, Virginia. Soft spoken and conscientious, Nguyen examined Richmond and found nothing exceptional. Like people everywhere by then, he knew about the Daschle anthrax letter, though he doubted it had anything to do with Richmond's illness. Still, he "had this feeling in the back of my neck," he told me, that prompted him to send Richmond to the hospital for further assessment.

After driving her husband north for 20 minutes from Nguyen's office, Susan Richmond turned off Route 495 onto Gallows Road in Falls Church. There, at about 5 p.m., she parked near the entrance to Inova Fairfax Hospital and helped her husband to the emergency room. Leroy Richmond's case was not deemed critical, and he had to wait until people with heart attacks and other clearly urgent problems had been seen. After 4 hours, a doctor appeared.

Cecele Murphy is a tall, ebullient blonde. A former flight attendant, she later became a professor of communications at Purdue University. Then in 1988, at age 37, she began another dramatic career change: she entered medical school. After her residency in emergency medicine at George Washington University, she started working at Inova Fairfax in 1996. That Friday afternoon, while tending to dozens of emergency room cases, she received a briefing about the patient in Cubicle 8. She had seen his chest X ray—mild pulmonary effusion—and heard he had some breathing difficulties. Dr. Murphy drew the curtain behind her as she entered the 7-foot-wide space. She smiled at Susan Richmond, who had not left her husband's side since their arrival at the hospital and turned to Mr. Richmond.

"Hi, how are you? I'm Dr. Murphy. The physician's assistant tells me you're not feeling well."

"I really know my body, and there's something wrong," he answered.

Susan chimed in: "He is never sick. I mean, if he says he's sick, he really is."

Murphy thought to herself, "He's not pale because he's dark skinned, but he looks and sounds very weak." Midway through the interview Richmond said, "I work at Brentwood."

"What's Brentwood?" Murphy asked. "It's the post office that handles mail for the Senate."

She thought about whether this revelation had meaning: "None of us knew anything about Brentwood at that point. We knew all about the Daschle letter and the Senate because we were getting reports from the CDC and the Washington Health Department about what to do with patients who come in and say they were on the fifth floor in the Hart Building. But about Brentwood, nothing."

She doubted that Richmond's illness was connected to the Daschle letter, but she felt she could not ignore the possibility. His chest X ray showed no mediastinal widening—a tell-tale sign of inhalation anthrax. But she wanted a more refined view, so she ordered a CT scan of his chest. "CT," or "CAT," stands for computerized axial tomography. It is an x-ray technique that produces multiple images of a part of the body taken from different angles. A computer reconstructs the images into vertical cross sections that can be viewed layer by layer, like slices of salami.

Meanwhile, Richmond's condition seemed to worsen as he lay in the emergency room. He had developed a fever and showed blood in his urine. Murphy drew blood samples for testing and then ordered him on multiple antibiotics, including intravenous Cipro. When the radiologist saw Richmond's CT results that night, he called Murphy: "What were you thinking you might find in this CT?" he asked. She mentioned pneumonia, cancer, maybe anthrax. He replied, "Well, he's got significant mediastinal lymphadenopathy. It looks like there's blood in his lymph nodes." Even though Murphy had thought of anthrax as a possibility, hearing that the lymph nodes were swollen and bloody stunned her. "Are you playing with me?" she asked. "No," the radiologist responded. She did a quick mental inventory: Brentwood, rapidly failing health, now the nodes. "My God, then he really does have it." She returned immediately to Cubicle 8. "Mr. Richmond, your CT shows swollen lymph nodes in your chest. That could mean anthrax." Richmond remembers hearing Dr. Murphy say "anthrax" but decided she must be wrong. It had to be pneumonia, he believed.

At 1 a.m Richmond was admitted to the intensive care unit. Around that time Dr. Susan Matcha received a phone call at home from the emergency room. "We have a patient here who could have anthrax." Matcha was the Kaiser plan infectious disease doctor covering three hospitals that weekend. She was told the patient's blood was being cultured, that he had a chest X ray, and that he was on Cipro. She answered: "Okay, it sounds like the appropriate things are being done. I'll see him in the morning."

The next morning, as she was driving to Inova Fairfax, her cell phone rang. It was from another physician at the hospital. "You better get over here right now. It's really looking like the patient has anthrax." Leroy Richmond's blood culture had grown organisms consistent with anthrax. She rushed to the hospital and examined Richmond. After phone consultations with the CDC, Matcha added rifampin and clindamycin to the Cipro, hoping that the additional antibiotics might help fight the organism and its toxin.

On Monday afternoon, 3 days after his admission, Richmond

received an unexpected visitor. Dr. Nguyen had come by, not to examine him but to show him something. "These are incense sticks," he said. Nguyen, a Buddhist, told Richmond that he would be going to his temple, where "I'll be burning the incense and praying for you." Richmond nodded and with effort managed a "Thank you." Nguyen became increasingly emotional. "I am so happy to see you alive and that you went to the emergency room on Friday." As Dr. Nguyen spoke, Richmond and his wife, Susan, saw tears in his eyes.

Meanwhile, on Saturday night, the day after Richmond went to the hospital, another man was admitted to Inova Fairfax with similar symptoms. Nine months later, Qieth McQue was still insisting on his anonymity, but for the first time agreed to speak about his experience. (In mid-2003 he also gave me permission to cite him by name, McQureerir, or by the short version that he commonly uses: McQue.)

Like Richmond, McQue belongs to the Kaiser health plan, and at 8:30 Saturday morning, October 20, he called to see a doctor. "They had nothing open until 3:30 in the afternoon, so that's when I went," he says. According to McQue, the doctor seemed puzzled: "He asked me was I kissing anybody? Was I drinking anything? I said, 'No,' Then as soon as I said 'Brentwood,' they admitted me."

When Dr. Susan Matcha, who was already treating Leroy Richmond, heard about McQue's symptoms and where he worked, she ordered that a blood sample be drawn, and then placed him on intravenous Cipro. She did not tell the patient what her suspicions were, pending the results of his blood test. Fifteen hours later it was clear: *Bacillus anthracis* had been cultured from his blood. Dr. Matcha came to Qieth McQue's room with a message that overwhelmed him:

> "Do you know what anthrax is?" McQue recalls her asking.
> "Hell no," he answered.
> "Do you want to hear the good news first or the bad news?"

McQue thought: "Well, the bad news might be that I have to stay here for 2 or 3 days. The good news would be that I can go home now with some pills."

"Let me hear the bad news first," he said.

"You have inhaled anthrax spores. It's a fatal disease. But we think we got it in time."

McQue, petrified, responded: "You've got to get out of this room, now." He explained:

> I didn't want to see anyone. Let me talk to my God, I thought. You know, you feel the rug has been pulled from under your feet. The IV tubes are in your arms and legs pumping things in you, and someone comes in and says you could be dying. I come from the West Indies where a hospital is just for two things: birth and death.

Matcha's recollection differed somewhat. She does not remember being asked to leave the room. "And I'm not really sure I said anything to the effect of 'I think we can help you,' because I wasn't so sure we could help at that point."

When Matcha visited McQue the next day, Monday, he was better able to talk. "But all I can think is, 'How can I get out of this hospital?'" McQue said. Matcha was also there to tell McQue and Richmond that she would no longer be treating them. She had been covering at Inova Fairfax over the weekend but would be returning to her base at the Washington Hospital Center, near the Capitol. Two infectious disease specialists at Inova, Jonathan Rosenthal and Naaz Fatteh, would be taking over their cases. From then on, Rosenthal assumed primary care for Richmond and Fatteh for McQue, though each doctor saw both patients and continuously discussed their progress with each other.

After medical school at Columbia University, Rosenthal took a residency at Case Western in Cleveland in the late 1980s. In 1991 he moved to Northern Virginia, outside Washington, D.C., to work at Kaiser Permanente. On Saturday morning Susan Matcha had called to tell him about Richmond's positive blood culture. "She called me because I'm based at Fairfax and I'm responsible with Dr. Fatteh for Inova Fairfax patients." Rosenthal had been fully briefed by the time he saw Richmond and Qieth McQue on Monday.

Rosenthal speaks quickly. His manner is direct, no nonsense. How did he feel about taking on the treatment of the two men? "Initially we were terrified that they were just gonna drop dead on us within a few days—because of the impression from the literature that if you get to this stage of the disease, the mortality is just so overwhelmingly high."

He spoke of the agony Richmond and McQue faced with every breath. "Both patients were rapidly developing bloody pleural effusions." The poisonous fluids were taking up more and more room in their chest cavities: "I took care of Leroy in particular. He had major lung compression, and he had to be drained repeatedly. You know, two thirds of his pleural cavity was filled with fluid, and his lung was compressed to a third of its previous size."

Toward the end of the week, after 2 quarts of fluid had been drained, the pressure on Richmond's chest suddenly eased. Months later, he called out "I can breathe!" as he recapitulated his moment of relief from the torment. "It was like a miracle. I can breathe!" In the course of the next few weeks, Richmond and McQue continued to improve, though Richmond suffered complications along the way, notably severe anemia. On November 9, Qieth McQue was deemed well enough to go home, and 5 days later, after 27 days in the hospital, Richmond was discharged.

Dr. Rosenthal thought about the fortunate sequence of events. After Qieth McQue had arrived in the emergency room on Saturday, his head scan and spinal tap "didn't show anything." But by then Leroy Richmond was known to have had anthrax. Rosenthal continued:

> Because Susan Matcha was on call that weekend, the internist called her from the emergency room and said "We have this guy [Qieth McQue] with a big headache. He doesn't seem to have meningitis. By the way, he works at the post office." And she said, "Okay, get a CAT scan." But if it had been the other way around, and he had come in first—before Leroy Richmond—he probably would have been sent home and called the next day. And maybe it would have been too late for him.

Rosenthal, like many of the doctors involved with the anthrax cases, was sought out by the news media. He saved his first interview for the newspaper that his father, A. M. Rosenthal, once edited, the *New York Times*.

Nearly a year later, neither Richmond nor McQue had fully recovered. Both remained at home much of the time, weak, tired, and suffering from short-term memory loss. But while Richmond seemed unresentful, McQue remained angry: "I feel like a pregnant moose," said McQue. He wakes up nauseated and tired and needs daytime naps. "I'm 54 and I feel like I'm 90. I have to walk with a cane." As difficult as their experiences have been, both men thank God for having survived. The outcomes for two of their co-workers turned out to be more tragic.

On Thursday, October 18, the day before Leroy Richmond

sought medical care, his former card partner, Thomas Morris, went to the doctor. Morris was also a member of the Kaiser plan and visited one of the plan's doctors in Maryland, near his home in Suitland. The doctor believed Morris's fever, sweats, and muscle aches were symptoms of a viral infection. He dismissed Morris's concerns that they might be related to anthrax and sent him home. Three days later, on Sunday, Morris dialed 911:

"What's the problem?" the operator asked, according to the taped transcript.

"My breathing is very, very labored," Morris answered.

"How old are you?"

"I'm 55. Ah, I don't know if I have been, but I suspect that I might have been exposed to anthrax."

The operator seemed flustered: "You know when or what—?"

"Ah, it was last what, last Saturday, a week ago last Saturday [October 13] morning at work. I work for the Postal Service. I've been to the doctor. I went to the doctor Thursday. He took a culture, but he never got back to me with the results. I guess there was some hangup over the weekend. I'm not sure. But in the meantime, I went through achiness and headachiness. This started Tuesday. Now I'm having difficulty breathing. And just to move any distance I feel like I'm going to pass out."

In another portion of the tape Morris amplifies on his suspicion that he may have been exposed to anthrax:

"It was—a woman found the envelope and I was in the vicinity. It had powder in it. They never let us know whether the thing had—was anthrax or not. They never treated the people who were around this particular individual and the supervisor who handled the envelope. So I don't know if it is or not. I'm just—I've never been able to find out. I've been calling. But the symptoms that I've had are what was described to me in a letter they put out, almost to the 'T,' except I haven't had any vomiting until just a few minutes ago. I'm not bleeding, and I don't have diarrhea. The doctor thought it was just a virus or something . . . So we went with that and I was taking Tylenol for the achiness. But the shortness of breath, now, I don't know. That's consistent with the—with anthrax."

"But you weren't the one that handled the envelope? It was someone else?"

"No, I didn't handle it. But I was in the vicinity."

"OK. And do you know what they did with the envelope at work?"

"I don't know anything. I don't know anything. I couldn't even find out if—if the stuff was or wasn't. I was told that it wasn't, but I have a tendency not to believe these people."

Following reports about Morris's 911 call, Deborah Willhite, a spokeswoman for the U.S. Postal Service, said that the FBI had tested the letter he mentioned and found no evidence of anthrax. Meanwhile, after the call, the operator dispatched an ambulance, which brought Morris to Greater Southeast Hospital in Washington. He arrived at 3 p.m. Less than 6 hours later, at 8:45 p.m., he was declared dead.

The following morning, Monday, October 22, Celeste Curseen also dialed 911 from her home in Clinton, Maryland. Two evenings earlier, her husband, Joe, had not been feeling well and had passed out while attending mass at St. John the Evangelist Catholic Church in Clinton. The next day, Sunday, he felt worse—tired, nauseated, perspiring beads of sweat "as big as half dollars," Mrs. Curseen said. She drove him to Southern Maryland Hospital in Clinton, where he was diagnosed with dehydration and gastroenteritis and sent home. Before dawn on Monday morning, she awoke to find him doubled over on the bathroom floor.

"He's breathing just constantly. He's got asthma, and he's just constantly breathing hard and fast," Celeste Curseen told the 911 operator.

"How long was it that he, um—?" the operator asked.

"I don't know. I fell asleep. I was asleep and I just looked up and he was laying out in the bathroom there."

"Is he able to talk to you normally?"

"No. He's breathing so hard. Sometimes he won't say anything for a period of time. But yes, he's talking."

"OK. Is he able to, say, talk in a complete sentence?"

"No, he's just been answering my questions."

Celeste helped her husband into the car and took him back to Southern Maryland Hospital. Dr. Venkat Mani, the hospital's head of infectious diseases, was called in. He reviewed the reports of Curseen's visit the previous day and saw nothing alarming in them. But now, after viewing Curseen's blood under the microscope, he knew there was little that could be done. His blood was teeming with anthrax bacilli: "There were so many organisms on the smear, that we could directly see it. When you have a person with blood infection, and you see the bacteria on the blood smear, the patient will almost never survive. By that stage the bacteria [are] winning the battle." Like his friend Mo, 6 hours after being admitted to the hospital, Joe Curseen was dead.

The day before, Sunday, when Curseen was brought to Southern Maryland Hospital, Leroy Richmond and Qieth McQue had already been hospitalized with suspected anthrax at Inova Fairfax. One week later, Dr. Ivan Walks, the health director for Washington, D.C., said that several hospitals in the area had been notified early about the anthrax cases at Inova. Southern Maryland was not among them. That hospital had been "outside of our perimeter for that symptom reporting," Walks said. No longer. "Since then we have learned a lesson, and we've expanded that perimeter now."

During the weekend of October 20-21, yet another mail handler in the area was feeling ill. "All I wanted to do on Saturday and Sunday was lay around and nap," recalled David Hose in his Virginia-accented baritone. The next day, Monday, he reported for work at the State Department annex near Dulles Airport, where he supervised the sorting of mail for delivery to State Department offices. He sweated through a sleepless Monday night at his home in Winchester, Virginia, but went to work the next day. By Tuesday night, "I was getting aches and pains all over. Really terrible." After a night of vomiting, he went with his wife to the emergency room at Winchester Medical Center. He had already begun to suspect he might have anthrax: "I mean I was really feeling strange. I've had the flu before. I've had bronchitis before. I've had pneumonia before. I've had all kinds of stuff, you know, years and years apart. But this felt like none of that."

The emergency room doctor doubted that he had anthrax, said Hose. But he drew a blood sample and gave Hose a Cipro pill to take just in case. He wrote prescriptions for more of the antibiotic and for cough syrup and sent him home. The next morning, Thursday, October 25, Hose felt so sick that he decided to call 911. "I needed the rescue squad." But at 7:55 a.m., as he was reaching for the phone, it began to ring. A voice announced that the results of his blood culture were in. "Mr. Hose, you definitely have anthrax. We'll send an ambulance for you. You've got to get in here now."

During his hospital stay, Hose repeatedly had fluid drained from his pleural cavity. At the same time, he developed bleeding ulcers and needed a blood transfusion and stomach cautery. On November 9, after 16 days of hospitalization, he was discharged. It was the same day that Qieth McQue was released from Inova Fairfax Hospital. Months later, like McQue and Richmond, Hose still felt sick. Although free of anthrax infection, like the other two men, he was weak, tired, and prone to memory lapses.

In August 2001, George DiFerdinando was appointed acting commissioner of New Jersey's Department of Health and Senior Services by Donald DeFrancesco, the acting governor. DeFerdinando will never forget his first staff meeting. "It began at 9 a.m. on September 11," he said, pausing over a cup of coffee and a muffin nearly a year later in a luncheonette in Princeton. Born and raised in New Jersey, DeFerdinando had lived outside the state for most of the previous two decades. After completing medical school and a residency in epidemiology at the University of North Carolina, he went to work for New York State's Department of Health. Then in 2000 he became deputy commissioner of health in New Jersey.

Under a shock of thick black hair, he offered a quick smile as he recalled that initial staff meeting on September 11. Some 60 people were in the room. "I'm hyperactive," he said, "and I'm always checking my BlackBerry." His pocket-sized communicator indicated something about "breaking news." Just then his secretary handed him a note saying two planes had hit the World Trade Center. "Wow," he said aloud and informed the group. "I told everyone to go back to their stations." In coordination with New York City officials, DiFerdinando led New Jersey's public health response. Thousands of New Jerseyans—survivors and families of victims—had been touched directly by the airplane homicides. But whereas New Jersey was working with New York City on the September 11 event, the anthrax outbreak a few weeks later placed the state at center stage.

"George," the caller said, "this is Polly Thomas." Dr. Thomas, a physician with New York City's Health Department, had known DiFerdinando when he was at the New York State Health Department. Her call came on October 12 at 12:30 p.m., about 9 hours after CDC Director Koplan's middle-of-the-night briefing to Marci

Layton. Thomas was on the line to alert him that New York City's health commissioner, Neal Cohen, and Mayor Rudy Giuliani were about to hold a press conference. "They're talking about a cutaneous anthrax case, an NBC employee. The health department is recommending antibiotics for certain people there," she said.

"Do you know the source of the bacteria?" DiFerdinando asked.

"There's confusion. It may be a September 25 letter. It might be one from September 18," she responded.

DiFerdinando thought to himself: "Okay, George, think clearly. Between one-fourth and one-third of the people at NBC are probably New Jerseyans. Some of them would go to Valley Hospital in Ridgewood, Holy Name in Teaneck, Hackensack Hospital, and others in the area. We've got to alert them all."

DiFerdinando and Thomas reminded each other that New York City's Health Department protocols for responding to bioterrorism were identical to New Jersey's. Area hospitals and health care providers needed to prepare for the worst. New Jersey's Office of Emergency Management is under the authority of the state police. "We immediately got them to activate the Emergency Operations Center phone bank," DiFerdinando says.

The next morning, Saturday, October 13, he met with six top health department officials. "Dr. Eddy Bresnitz, the state epidemiologist was there, James Blumenstock, an assistant commissioner, others." Even though the anthrax letter sent to Senator Daschle would not be discovered for another 2 days, the New York case had already shocked everyone—it was the first outside Florida. They hooked into a conference call for an update from Health and Human Services Secretary Tommy Thompson and the CDC's Jeffrey Koplan. "I think every department of health in the United States was on the call," DiFerdinando says.

But an even more significant item of news had begun to circulate. Headlined "Anthrax Found in NBC News Aide," a *New York Times* article reported that the aide, Erin O'Connor, may have been infected by an anthrax-filled letter some weeks earlier. After developing a rash on her chest, the article said, the sore "turned almost black, and she developed a low-grade fever." DiFerdinando says: "The report got the attention of two physicians in New Jersey who were taking care of postal workers from the same post office who had unusual cutaneous illnesses. Each doctor called in about their patient. Both patients had become ill within a few days of each other in late September."

The FBI interviewed the physicians and found that one of them had sent a patient to a plastic surgeon who had removed the infected tissue and preserved it. When the FBI learned about this, the bureau had the specimen sent over to the state laboratories.

Dr. Faruk Presswalla has been New Jersey's chief medical examiner since 1997. After immigrating to the United States from India in the 1960s, he worked in the health departments of New York City and Virginia. "I was reading about the first cutaneous case, Tom Brokaw's assistant," he recalled, "when my phone rang." It was Dr. Eddy Bresnitz, New Jersey's epidemiologist. In some states the office of the medical examiner is under the state's health department but not in New Jersey. There it is under the Department of Justice. Bresnitz was calling from the Department of Health:

"Faruk, we have a biopsy from a postal worker who had a skin lesion. Her doctor remembered it was black and didn't resolve—had to be removed by a plastic surgeon. We'd like you to look at it for anthrax."

"Why don't you send it to CDC?" Presswalla asked.

"The CDC said they couldn't do it quickly. They're being bombarded with requests for tests. The FBI is anxious."

Presswalla hesitated. He told Bresnitz that his laboratory was not equipped to provide a confirmation of anthrax. "But I can do the histology and rule out a spider or arthropod bite." If the lesion was caused by a bite, Presswalla said, a microscopic examination should show evidence of a blister or redness along with leukocytes—white blood cells—that the body dispatches to destroy disease-causing organisms.

Late in the evening of October 15, the FBI brought the specimen to Presswalla's laboratory in Newark. The next morning, he prepared it for examination. Meanwhile he remembered that some 40 years earlier in India he had seen a case of cutaneous anthrax. Presswalla is short and his dark hair has thinned out. In a melodious East Asian accent, he explained: "My mind went back to that time. Now, when I looked through the microscope, it seemed familiar—a paucity of leukocytes, local necrosis, Gram-positive rods. *Bacillus subtilis* and *Bacillus cereus* look like rods, but in that context—in tissue—I was 95 percent certain that it was anthrax."

He called Dr. Bresnitz and the FBI and provided them with the specimen and slides for delivery to the CDC. He contemplated what all this could mean and began to worry about his own mail: "I told people in the office, when the mail comes in, we wear gloves and wipe everything with a wet paper towel. And I told my wife not to open mail. I'll do it. I wore gloves and I wiped envelopes and opened them on the hardwood floor at home."

Presswalla's worries during the week of October 15 mirrored much of the nation's. Another doctor who had treated an anthrax patient told me that he and other infectious disease specialists feared they would be targeted by bioterrorists. Reports of anthrax in Washington, New York, and now New Jersey seemed to be rolling out in an indecipherable jumble. Realizing that the mail was involved only added to the anxiety. How many anthrax letters had been sent? How dangerous was other mail that the contaminated letters may have touched? How dangerous was the post office? These uncertainties played on people's minds everywhere. Senators, doctors, teachers, laborers, police, firemen, FBI agents—all get mail.

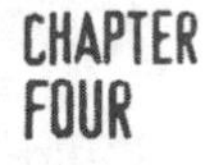

ULTIMATE DELIVERY: THE U.S. MAIL

On Friday night, October 19, 2001, Joseph Curseen went to work for the last time. Three days later he was dead. For 15 years Joe had been pulling into the parking lot of the mammoth postal distribution center on Brentwood Road in northeast Washington, D.C. He was there when the facility opened in 1986 and a year later when rows of rectangular holes mysteriously began to appear in the parking lot. "Brentwood," as the center was called, had been built over a graveyard and the asphalt was sinking into empty graves.

The 632,000-square-foot structure is large enough to accommodate 10 football fields. Until it closed because of anthrax contamination in 2001, its machines whirred and hummed through three shifts, 24 hours a day. With the help of these machines, 2,100 employees toiled in its cavernous space to separate, sort, and deliver nearly 2 million pieces of mail every day.

Joe Curseen was part of that routine. Upon arriving at work shortly before the 10:30 p.m. to 7 a.m. shift, he would hang his coat in the locker next to the cafeteria on the second floor. An escalator ride to the vast ground floor area would bring him near the time clock. That Friday, as always, he met his work partner, Jim Harper, next to the clock. And as always, Joe shook his hand and asked, "How you doing, Harper?" It was a ritual of friendship that Harper much appreciated.

After punching the clock, they turned right and walked past an array of cubicles and machines. In every direction stood racks, white plastic tubs, carts of assorted sizes, some containing mail to be processed, some empty and awaiting incoming shipments. Broad conveyor belts, angled at 30 degrees, were moving unprocessed mail up into chutes for initial separation. Yellow lights blinked in synchrony with loud beeps to signal that a sorting machine was being turned on. The feeling was that of an amusement park in which the mail was being treated to dizzying rides and bright lights.

Toward the far end of the building, Curseen and Harper arrived at machine number 17, the "delivery-bar-code sorter" that they had been operating together for 2 years. The machine contained four rows of bins, one above the other, extending more than 60 feet. Altogether there were 250 bins.

"Joe, you want to start loading and I'll sweep?" Harper asked. That was the way they usually began their shift. Sweeping, in this case, meant pulling the sorted mail out of the bins. "Sure," Curseen answered, knowing that in an hour he and Harper would switch positions as a matter of course.

Curseen felt as if he was coming down with a cold; nothing serious. He reached into the cart next to him, wedged about 200 letters between his palms, and placed them on the waist-high conveyor belt that feeds into the machine. Bill Lewis, a veteran postal mechanic, knows the sorting machines inside out:

> The machine has rollers that pull in one letter at a time—we call it "shingling." There's such a tight fit that the letter becomes compressed. By the time a letter gets through the process with those tight belts and rollers, there's no air left in the envelope. It's been squeezed out.

How long does it take for a letter to go through a sorting machine? "It goes so fast you can hardly follow it with your eyes," Lewis said. Three or four seconds from beginning to end, including a process that sucks the letter under an electric eye that reads the bar code and directs it to a bin.

Curseen's sorter was the last machine through which a letter would pass. By that time other machines had already imprinted cancellation lines and bar codes on the envelope. But the ride through the sorting machine was also quick and turbulent. Letters poured into the bins at a rate of 13,500 per hour. Harper's job, alternating with Curseen's, was to pluck them from the bins and place them in plastic tubs. The sorted mail would be loaded on trucks headed for post offices, perhaps to an airport, or directly for local delivery.

Appreciation of the mail as an essential means of communication in America predates the birth of the republic. In 1775 the Continental Congress named Benjamin Franklin, perhaps the most venerated citizen of the time, as the first postmaster general. (Historian Alan Taylor deemed Franklin "the leading figure in colonial politics, literature, science, and social reform.") His mandate included the establishment and regulation of post offices "from one State to another." After the Constitution was adopted in 1789, one of the first acts of the new Congress was to establish the Office of the Postmaster General. The office was given authority over the country's 75 post offices and its 2,000 miles of post roads and was seen as instrumental in fostering both commerce and national unity.

During the next two centuries, the operations of the postal service were linked to the advance of technology. Transportation of the mail, which started by horse, later included conveyance by rail, sea, and air. To this day, mail is delivered by train, truck, boat, and airplane, and in the Grand Canyon by mule. Despite the emergence of other methods of communication—notably the telephone and the Internet—postal communication remains a vital part of American life.

In 38,000 post offices, the U.S. Postal Service handles more than 200 billion pieces of mail a year. Every person in the United States receives, on average, two to three mail items a day. In every part of the country the mail is channeled through large distribution centers like the one on Brentwood Road in Washington, D.C.

Thursday, October 18, 2001, was a pivotal day in the anthrax crisis. Although feeling out of sorts, Joe Curseen was still working at his post with his friend Harper. Until that day, four people were known to have been infected with anthrax: Bob Stevens (who had died) and Ernie Blanco from American Media, Inc.; Erin O'Connor from NBC; and the baby of an ABC employee in New York City. In addition, the day before, on October 17, twenty-eight people were found to have been exposed to anthrax by the letter sent to Senator Daschle. In all these cases the risk seemed limited to people in the vicinity of an opened letter. On October 18, another such case, that of Claire Fletcher, who opened mail at CBS in New York, was reported.

But additional findings that day raised new and larger worries.

A lesion on mail carrier Teresa Heller's arm was confirmed by the CDC as the first anthrax infection in New Jersey and one on mechanic Richard Morgano's was a suspected second. (His blood specimen contained antibodies to anthrax.) Moreover, they were the first postal workers identified as having anthrax. Heller was based in the West Trenton office in Ewing Township and Morgano at the nearby Hamilton distribution center, where letters were postmarked "Trenton, NJ."

October 18 was also a day of paradox. Although postal employees had not previously been considered at risk, many were worried, especially at the Brentwood and Hamilton facilities. To dispel anxiety, U.S. Postmaster General John Potter decided to make an appearance at Brentwood. On Thursday afternoon, October 18, he visited the center and mingled with employees. At a press conference in the large working area, he announced a $1 million reward for information leading to the culprits responsible for the anthrax letters. Potter assured the assembled workers that they were not in danger. The Daschle letter, he said, "was extremely well sealed, and there is only a minute chance that anthrax spores escaped from it into the facility." Neither Potter nor anyone else realized that, as he was speaking, postal workers there and elsewhere had already been infected.

Officials did not order the Brentwood facility closed until Sunday, October 21. But in New Jersey the decision to close Hamilton came earlier. It was prompted by the news about Teresa Heller and Richard Morgano. Heller had first noticed a red sore an inch above her wrist, on the palm side, on Friday, September 28. "It itched a lot," she said "and in a couple of days it blistered and turned into a brown scab."

Heller, 45, had been a mail carrier for 8 years at the West Trenton post office. Soft spoken and reticent, she has routinely refused requests for media interviews. One year after the incident, she was willing to tell me about her experience. On October 1 she had decided to see her physician, Dr. Elsie Jones. (The name is a pseudonym at the doctor's request.) She said, "By the time I went to the doctor, the scab had gotten bigger and turned black, and my wrist had started to swell up. I told her that someone at work said it looked like it could be a spider bite."

Dr. Jones clearly remembers Heller's visit. "The lesion looked deep and purplish. I had never seen anything like it before," she said. "Were you doing anything unusual, maybe outdoors?" Dr. Jones asked Teresa. Teresa couldn't think of anything out of the ordinary. "You know, when I'm walking on my route, sometimes I go by some bushes." "Any insects bite you?" "Not that I remember."

They both understood, of course, that Teresa could have been bitten and not realized it. Elsie Jones, a family medicine practitioner, prescribed Avelox, an antibiotic that, like Cipro, is in the quinolone family. Because the infection seemed like it might have reached her bone—"it was the deepest cellulitis I ever saw"—she made an appointment for Teresa to see an orthopedist, thinking he would scrape out the necrotic tissue.

That night Teresa felt worse. "I started getting a fever, and my arm was swelling from my wrist to my elbow." The next day the orthopedist looked at her arm and said he could not do anything for it. Then he and Dr. Jones decided to send Teresa to a plastic surgeon.

By the time Dr. Parvaiz Malik, the plastic surgeon, saw Teresa on October 3 she had "spiked a fever" and he put her in the hospital. On October 5, Teresa was discharged. That was also the day Bob Stevens died in Florida, though at that point no one dreamt that the two cases were related. Meanwhile, Dr. Malik had scraped out the blackened wound on Heller's arm. "Most of the time I just throw necrotic tissue away," he says. "But this was a good size, and I tossed it in a bottle of formalin and left it sitting in my office." Formalin, a mixture of formaldehyde and water, is a preservative as well as a germ killer.

Dr. Jones's husband, also a physician, shares an office with his wife. He saw Teresa's lesion when his wife did, and he was equally puzzled. On Saturday morning, October 13, the day after the CDC's confirmation of the first skin anthrax case in New York, Dr. Jones was out shopping with her children. Her husband was at home, listening to the news on the radio. The first thing he said to his wife when she returned was: "What if I told you there was a cutaneous anthrax case at NBC in New York?" She understood the implications immediately. "A light flashed on for both of us."

Dr. Jones had not only been Teresa's physician for 10 years, she was also a neighbor. Their homes are in Bordentown, and "she lives right around the corner from me," Teresa said. That Saturday afternoon Dr. Jones "stopped by my house. She told me what she had just heard, and that she had been reading more about the symptoms of anthrax, and that they sounded like mine. She knew I was a mail carrier. She said, 'It just seems like too much of a coincidence.' Then she asked if I wouldn't mind if she contacted the authorities." Teresa, whose husband is a sheriff's officer, readily agreed. Even though she was feeling better by then, when she heard she might have anthrax, she was "a little shaky."

In thinking about that afternoon, Dr. Jones evinced some frustration. She returned home and called the local police, "but they weren't interested." Then she tried the state police. "They weren't interested either, but they suggested we contact the FBI [Federal Bureau of Investigation] or the state's emergency hotline. I called both."

She also phoned Dr. Malik. When his pager went off, Malik was at a fundraising dinner for his hospital, the RWJ Hamilton. He left the dining table to return the call. "It looks like Teresa Heller's lesion might be anthrax," Dr. Jones said. "I'm shocked," he replied, and acknowledged later that he felt frightened by the news. When he returned to the table his wife asked, "What's wrong? You look troubled." He told her what he had just heard and then thought about the tissue specimen that was still in his office. "I called the CDC and said someone should pick it up and look at it. They suggested that I call the FBI." When Malik reached the FBI, an agent said he would meet him at his office.

That night at 10 p.m., as Malik handed over the bottle containing Teresa's black specimen to the FBI agent, he said, "I hope you can help get this checked quickly." "Yes, we can do that," the agent said. The specimen was delivered to Eddy Bresnitz, the state epidemiologist, then to Faruk Presswalla, the medical examiner, for initial testing, and ultimately to the CDC.

Malik did not hear from the FBI or any other authorities. Five days later, while watching television, he learned that the skin specimen had tested positive for anthrax. He was eager to see the laboratory report. "I called the FBI twice and then the CDC." Months later he still had not seen the report. "I took pictures of the lesion, and I wanted to write an article for a medical journal, but I can't do it until I get the report."

Meanwhile, Teresa continued to improve, but the wound in her arm remained deep and unsightly. Dr. Malik grafted some skin over the area, which helped with its appearance. "When the skin graft healed up, I felt a lot better," Teresa said, and soon after she decided to go back to work. The day she returned was a day of quiet celebration. "It was a week before Thanksgiving, on November 17." A year later she said "I feel fine. I still go on the same mail route in Ewing Township."

Did she ever think about leaving her job because of the experience? With a chuckle she answered, "No, I just think it was a one-shot thing. I just keep going." Still, the circle of scar tissue on her forearm—nearly the size of a half-dollar—will never let her forget about her difficult experience.

The Saturday night that Malik gave the specimen to an FBI agent two other agents went to Dr. Jones's home to discuss Teresa Heller's case. They also asked if Jones would be willing to look at Richard Morgano, the second postal worker they had heard about that day. The next day the agents brought Morgano to her house. Dr. Jones examined his right arm and said, "It looks like it to me." Morgano was sure that he had become infected at the Hamilton center. "They're saying that the building is safe, and I'm saying there's no way the building is safe. I know there's anthrax in there."

Eddy Bresnitz became New Jersey's chief epidemiologist in 1999. A native of Montreal who attended medical school at McGill University, he came to the United States in 1974 for training in epidemiology. He later taught at the Medical College of Pennsylvania where he became chairman of the Department of Community Medicine. Three years into his job at the New Jersey health department, he was in the thick of the anthrax crisis. He was the Health Department's contact with Dr. Faruk Presswalla, to whom he had sent Teresa Heller's biopsy. On October 16, Dr. Presswalla told Bresnitz by phone what he later formalized in a memo: "In my opinion, the gross specimen and microscopic examination are highly suggestive of cutaneous anthrax." The required staining techniques for a definitive diagnosis were not available in his laboratory. Bresnitz then made arrangements for the specimen to be sent to the CDC. On Thursday, October 18, at 7:30 in the morning, just before leaving home, he received a call from Robert Kinnard of the CDC:

"Ed, it's confirmed," Kinnard said.

"Are you sure about this?"

"We're absolutely sure."

"Oh boy. That is something," Bresnitz answered, and thought to himself, "This is going to be a busy day."

He thought about his visit, 3 days earlier, to the Hamilton postal center. Some 400 workers had assembled in the cafeteria to hear him. He told them they were basically not at risk from the anthrax letters that had been processed there. "The envelopes were sealed with tape," he said, "and there's been no evidence of anthrax illness here." Now, however, with the news of Teresa Heller's anthrax and the suspicion of Richard Morgano's, Bresnitz returned to the center. "It was 11:30 a.m. when I got there, and no one was working. The sorting machines were not running." Again, he met with the workers in the cafeteria and tried to answer their questions. It was clear to Bresnitz that "they were not going back to work." They needed no convincing when he told them the building would be closed. When he went outside, he unexpectedly saw Congressman Chris Smith and several Postal Service officials meeting with the press. Their message to reporters: "This place is closed for now. We're going to be doing more testing."

That afternoon, six CDC officials were en route to New Jersey by air from Atlanta, due to arrive by evening. They would be joining in a meeting with state health department staff and postal officials to review everything known about the situation. The meeting began at 9 p.m. at the Nassau Inn in Princeton. "There were about 20 people in the room and a few others from CDC on the phone," recalled George DiFerdinando, who, as the state's acting health commissioner, presided. "We talked about what we needed to do over the next 10 hours and the next day or two."

A difference of opinion emerged about when the Hamilton Center could be reopened. "We've hired a contractor who claims he can clean it in a brief period of time, somewhere between 24 and 48 hours," a postal representative said. "Well, how is he going to do this? What is his technique?" others wanted to know. "We're not really sure," came the answer. Still, the postal people seemed willing to rely on the word of the contractor.

The pressure to reopen was driven in part by the huge loss of dollars and efficiency if mail had to be processed elsewhere. Moreover, the fact that the Brentwood facility was still open also prompted some second-guessing. "What do they know in Brentwood that we don't know, because they're not closing?" one participant wondered. In the course of the evening, the postal representatives left the room a number of times to confer by phone with higher-level management. Annoyed by the interruptions, DiFerdinando eventually turned to one of the postal people and said: "If I can't negotiate with you, maybe we need to drive over to Hamilton where I can talk to the guys I'm negotiating with." "No, no, no, there's no need for that," came the answer.

Jennita Reefhuis, one of the CDC officials present, felt bad for

Dave Bowers and the other "postal guys," as she called them. "They were so torn, because they saw our point about the need for safety, but they also had a business to run and knew that people were depending on their mail."

Christina Tan, another CDC official, described the discussions as "confusing." The following year she joined the New Jersey Health Department as director of its Communicable Diseases Service. As we talked in her office, in Hamilton Township, I noticed her gold loop earrings and the large Chinese-language character hanging from her necklace. I asked if the character had special meaning, perhaps good luck. She laughed, "It's my name, Tan." On the wall to the left of her desk hung an eerily beautiful night photo of the Twin Towers, their windows brightly lit. In the same frame was a letter signed by Acting Governor Donald DeFrancesco, thanking Dr. Tan for her "extraordinary humanitarian work" during the terror crisis. As she sat below the letter, she thought about the extended to-and-fro of the October 18 meeting: "Whatever recommendation was made, you know, one way or the other, you could see the arguments for either side. Whether you opened the postal building or whether you closed it, at that time, you were damned if you do and damned if you don't, either way."

The issue of whether and when to reopen dragged on. "We weren't really getting anywhere," DiFerdinando said, as he recounted what happened in the end:

> It sounds a little silly, but it had been a long day, and around 1 a.m. I dozed off. You know, I was leading the meeting, and everyone was polite enough not to shake me. When I woke up, people were still debating whether to reopen the facility or keep it closed. I guess I got a little frustrated. I stopped the discussion by raising my voice and said I had figured something out. Whatever the postal service decided to do or not to do, I would not be standing there next to them if they reopened that facility.

The room fell silent. Then one of the postal representatives said, "Well if you weren't standing there, we wouldn't be able to open." "That's your decision," DiFerdinando answered. "I don't know what's going on in that building, so I can't vouch for its safety." That comment was the turning point. Minutes later everyone at the table came to an agreement. The facility would stay closed "until we knew more about the extent of contamination and how to clean it up."

Eddy Bresnitz was at that meeting. Ten months later, sitting in his office at the New Jersey State Department of Health and Senior Services, he considered what subsequently was learned about anthrax in the Hamilton facility. Wearing a dark business suit, his lean frame suggested that of a middle-aged distance runner. "Look at those dots," he said, pointing to the wall behind his desk. Hanging next to a map of New Jersey was a 3 × 5 foot diagram labeled: Postal and Distribution Center in Hamilton. The red dots were as pervasive as on a face full of chicken pox. Each represented an area where in subsequent months anthrax spores were found. No section of the 281,000 square foot structure was spared. How many dots are there? "I don't know the exact number, but it's hundreds and hundreds," Bresnitz answered. He drove home the point: "Even though there were just four anthrax letters [that we know of], this place is completely contaminated. Four corners. And what you don't see on the diagram is the other dimensions, from floor to ceiling, and in the ventilation system."

Bresnitz thought back to the meeting on the night of October 18 and the naïve optimism of the postal representatives. In the end, happily, they joined the consensus, announced by DiFerdinando, to keep the building shut. The diagram on the wall made clear that to have done otherwise could have been disastrous. On October 19 another Hamilton postal worker, Patrick O'Donnell, was found to have had skin anthrax. Again, DiFerdinando took the lead. The New Jersey Department of Health would now recommend that the 800 people who worked at the facility, and the 400 who worked at the local post offices it serviced, begin preventive antibiotic treatment.

On November 2, the *Wall Street Journal* ran an article titled "Seven Days in October Spotlight Weakness of Bioterrorism Response." It included a review of the anthrax incidents between October 15 and 22 that ended with the deaths of Thomas Morris and Joseph Curseen. Ricardo Custodio, a public health doctor, summarized the article in *Medical Editor's Column*, an on-line publication of health management groups, and identified a single bright spot during that devastating week:

> At least one unsung medical hero emerges from this tragedy, Dr. DiFerdinando. He ignored CDC advice and gave the New Jersey postal workers Cipro. He took action with little information and remained answerable for his decision. His concern for safety, patient-centeredness and timeliness possibly saved lives.

It was mid-summer 2002, and the sun hung high over central New Jersey. Two towns, 25 miles apart, shared the noonday brightness. Willingboro is more rustic and Princeton Junction is more affluent, but each is a picture of suburban calm. In both communities, clusters of comfortable single-family homes line quiet streets. And in both the residents reflect a relaxed diversity. They include hard-working families of many backgrounds—white, black, Asian, immigrant, and native born.

Elbow Lane in Willingboro leads to a winding street where branches, heavy with maple leaves, dip over rooftops. Set back from the inner curve of the street is a brick-and-wood two-story home. Norma Wallace smiled broadly: "I hope it was not too difficult finding your way?" Not bad, only one wrong turn, I answered. Norma's black hair hung in short thin strands. Her face was expressive. She sat at one end of a cream-colored couch near a large photograph taken a few months earlier, after she left the hospital. She pointed to each face: "That's my mother, my daughter, my granddaughter." A white gentleman was at their right. "That's Dr. Topiel, a wonderful man." Martin Topiel had led the team of doctors toward her recovery, such as it is. Nine months after contracting inhalation anthrax, she remained weak and had difficulty with her memory.

Norma was dressed in khaki slacks and an aqua T-shirt emblazoned "FLEX Gym." "I used to go there and work out," she said. "That was before I got sick. Now I do my exercising at Virtua." Virtua Memorial Hospital in Mount Holly was also where she spent 18 days before being discharged on November 5, 2001.

Norma began feeling ill on Sunday, October 14. "I felt like I had a cold. I was running a fever, vomiting, and I had diarrhea. I didn't know what I was experiencing, but I knew that I was sick." She took aspirin, drank a lot of water, and went to work on Monday and Tuesday, at the postal distribution center in nearby Hamilton Township. Her diarrhea and vomiting improved, but her breathing had become more difficult. "I went to the doctor on Wednesday, and he told me to take Tylenol because it would be more effective than aspirin." Her fever persisted. With the onset of chest pain, on Friday afternoon, October 19, she went to the Virtua emergency room.

Dr. Topiel, an infectious disease specialist, was walking out the door of the hospital when he was paged by the emergency room doctor. "We have somebody here who I think you need to see. She's from the post office and she's short of breath."

His white hair and mustache notwithstanding, Topiel looks and sounds like Richard Dreyfuss, the actor. "It was around 5 o'clock," he said. "She was lying on a stretcher and looking very ill, didn't seem like routine pneumonia to me." Like most people, he was aware of the anthrax outbreaks in Florida and New York. As she listed her symptoms, he thought about the possibility that she had inhalation anthrax. The doctor who had seen Norma 2 days earlier had treated her illness as a viral syndrome and had not prescribed antibiotics.

> I'm telling her, "It appears you have pneumonia," because I didn't want to raise her anxiety level. But I'm saying to myself, "This could be anthrax." I had her started on tests and treatment with the greatest rapidity that I've ever had performed. I called over two nurses and told them we needed to get an X ray immediately, to do a blood culture immediately, start her on antibiotics immediately. Everything was done in 20 minutes.

When Topiel mentioned his suspicions to some associates, they were skeptical. "Marty, don't be silly. This isn't New York. This is Mount Holly," one of his fellow physicians said. The reaction was reminiscent of the one heard by Dr. Larry Bush 2 weeks earlier in Florida. Like Bush, Topiel was persistent. "I think it's anthrax until proven otherwise," he said. When I told Dr. Topiel that Dr. Bush had used the same words—"anthrax until proven otherwise"—he grinned. "Larry Bush worked with me at this hospital before he went to Florida." They had rarely seen each other since. He thought for a moment and wondered aloud, "What are the odds that the two of us would have seen the rarest of illnesses in a bioterrorist event?"

Over the weekend Topiel was on the phone with a number of physicians—the covering doctor, the radiologist, the lung specialist, people at the New Jersey Department of Health. Late Saturday, when a second chest X ray showed enlarging pleural effusions, he and the pulmonary doctor decided that Norma's chest should be aspirated. On Sunday morning, on the way to a Jets football game with his 11-year-old son, Ben, Topiel stopped at the hospital. "I walked in when Norma was being tapped and I saw this bloody fluid being removed. My jaw dropped." He felt all the more strongly that he was seeing anthrax. But there had been no confirmation yet from the CDC. During the football game, Topiel was continuously on his cell phone, wondering about the blood culture taken on Friday. In recalling that time, he tells of an experience he had 20 years earlier.

Following medical school at New York University, in 1979 Topiel began 5 years of training in infectious diseases at George Washington University in Washington, D.C. In 1981, after the assassination attempt on President Reagan, Topiel assisted in caring for James Brady, his press secretary, who also was wounded. He recalled: "What I learned then was that there is something called 'V.I.P. medicine.' That means that when you do things differently than you normally do in routine care, mistakes happen. Incorrect care is sometimes given."

Norma Wallace had been fast tracked and given priority care—in Topiel's vernacular, V.I.P. medicine. On Friday night, after Norma's blood was drawn, he had contacted the state health department to say that her blood needed to be cultured for anthrax. Two days later, with no word about the results, he found out that neither the state nor the CDC had received her samples. Topiel's voice lowered, reflecting his continued sense of disappointment: "Apparently the blood samples that were taken somehow remained in the emergency room, because everybody was running around. So we didn't have the blood cultures to guide us at that point. That was the one time this happened in the 18 years I've been here."

When the CDC finally obtained the samples, no anthrax could be cultured. Whether the negative finding was somehow a result of the blood sitting on the shelf for 2 days is a question Topiel still ponders. Fortunately, the delayed testing only affected the timing of CDC's confirmation, not the treatment Norma received. She was already on intravenous Cipro, rifampin, and vancomycin and having fluids drained from her chest. It was only later, from tests conducted on her pleural fluid, that anthrax was confirmed. In any event, a week after her admission to the hospital, she was able to breathe more easily, and she began to feel stronger.

Just before being discharged, on the evening of November 5, Norma agreed to appear at a press conference at the hospital. Topiel walked into the conference room with her. "About 40 people were there," he recalled. "I did a little speech, very short. Then she went up to the podium and gave an incredible, powerful message." Norma recalls that evening clearly. What did she say? "We must not be fearful of the terrorists. We need to stand up against them. I just need to express that and not sit back about it." Although still feeling ill, Norma felt emboldened by her experience. But the effect on a fellow worker at the Hamilton postal center was very different.

A turn onto Lanwin Boulevard in Princeton Junction leads to a stretch of spacious homes and manicured lawns. As I drove slowly down the street, a cyclist pedaled sluggishly along the roadside path. Two streets in from the boulevard, a white colonial-style house is set behind a row of perfectly trimmed hedges. Jyotsna Patel, like Norma Wallace, had been a postal worker at the Hamilton facility. Wearing a loose-fitting maroon blouse and black pants, she smiled weakly. "Please sit down," she said, pointing to a firm white living-room couch. Her speech, though slow, carried the rhythmic sounds of her native India. She carefully lowered herself into a high-back chair at a right angle to the couch. Behind her the dining room was visible, eight chairs spaced evenly around a dark wood table. At Jyotsna's right a small table supported a photograph of her and her husband, Ramesh, an architect with a construction company in New York City. They moved to the United States in 1980, shortly after their marriage in India. "The picture was taken 4 years ago," Jyotsna said, when she was 39 and he was 43. Both faces appeared strong and vigorous. Until the fall of 2001, those images mirrored reality.

Dr. Baksh Patel (no relation to Jyotsna and Ramesh) thought back to Tuesday, October 16, 2001, when Jyotsna visited his Hamilton office with chest pain, shortness of breath, and fatigue. As her primary care physician, Patel had known Jyotsna to be focused and energetic. "Bronchitis," he decided. "Don't worry. Go home and get some rest, and everything will be fine." He prescribed levofloxacin, whose brand name is Levaquin. Like Cipro, Levaquin is in a class of antibiotics known as quinolones, which are effective against many of the same bacteria.

In our interview, Dr. Patel said, "Don't ask me why I gave her Levaquin." He realized of course that a different choice could have been useless against *Bacillus anthracis*, the bug later determined to be the cause of Jyotsna's illness. In short-sleeved khaki shirt and gold-rimmed glasses, he looked past the pile of charts on his desk. "It was a gut feeling that I can't explain." His satisfaction about his choice of drug was obvious.

Despite the prescription, the next day Jyotsna's condition worsened. On Wednesday evening Ramesh became frightened: "Around midnight she started to choke, couldn't breathe. When she talked she was confused. She was making no sense." After 20 minutes she recovered somewhat, and they decided to wait until morning to contact the doctor. When Ramesh called on Thursday, Dr. Patel said Jyotsna should stop at Capital Imaging, a radiology center near his office, to have a chest X ray taken and then to bring it to him.

By the time Jyotsna arrived at Dr. Patel's office on Klockner Road, it was 3 p.m. She could hardly stand. When the doctor looked at her X ray, Ramesh recalled, "he was really worried." Dr. Patel told them, "Her chest is filled with fluid. She has a real bad pneumonia." He turned to Jyotsna: "I'm calling the hospital. They'll be making arrangements for you, so go there right now."

That was on Thursday, October 18. By then, it was public knowledge that anthrax letters had been processed at the Hamilton postal facility. "I work on a sorting machine there," Jyotsna had told the doctor, expressing her concern that the letters might have something to do with her illness. The notion seemed far-fetched to Dr. Patel. And why shouldn't it? Health officials had not considered postal workers to be at risk from the letters. Nor could he have known that on that same day the CDC would confirm that another New Jersey postal worker had cutaneous anthrax. Teresa Heller, the woman with the skin anthrax, had nearly recovered by the time her diagnosis was made. Still, the implications were ominous. Heller was a mail carrier based in the West Trenton post office, which received mail directly from the Hamilton center on Route 130.

When Dr. Patel saw the white mass on Jyotsna's chest X ray, he thought it might be a malignancy. He knew there was a strong history of cancer in her family. Still, he was leaning toward pneumonia, certainly not anthrax. The news about Heller had not yet reached the public.

Jyotsna went directly from Dr. Patel's office to the Robert Wood Johnson Hospital in Hamilton, a half mile away. Ramesh and Jyotsna continued to worry about the diagnosis. "At the hospital, I insisted they check her out for anthrax," Ramesh said. Her nostrils were swabbed and a blood sample was placed in a nutrient growth medium, but neither sample grew *Bacillus anthracis*. As Ramesh later learned, the antibiotics she had been taking for 48 hours could have killed the bacteria, while their toxin continued to poison her body. He recalled that even 3 days after she had been admitted to the hospital, "they were still saying, 'pneumonia, pneumonia.'" In fact, her treatment would have been much the same had the diagnosis been anthrax. Jyotsna was now receiving intravenous Cipro and azithromycin, and fluids were being drained from her chest.

Days before Jyotsna entered the hospital on October 18, newly discovered anthrax infections in New York City were being reported by the city's health department. Then on the day that Jyotsna was admitted came the word of the first New Jersey cases, Teresa Heller and Richard Morgano, and on the next day a third case, Patrick O'Donnell. All were postal workers. This prompted DeFerdinando's decision to call in all 1,200 workers connected with Hamilton for testing and to start them on Cipro.

Ramesh had previously informed Jyotsna's supervisors that she was in the hospital with pneumonia. But on Saturday he received a call from postal authorities wanting to know why Jyotsna had not reported to the RJW-Hamilton Hospital for testing and to receive Cipro. Ramesh replied with a mixture of anger and amusement: "What do you mean she hasn't reported? She's been living in that hospital. Same hospital. She's in Room 21, and she's been there for 3 days."

After a week in the hospital, Jyotsna's symptoms had improved, and on October 26 she was discharged. Meanwhile, as the anthrax outbreak had become more clearly related to the mail, Jyotsna's diagnosis came under scrutiny. Even though her nasal swab and pleural fluid tested negative for anthrax, her symptoms seemed similar to the other inhalation victims. On the last day of her hospital stay, the CDC reported that the immunohistochemical staining was positive for samples from her pleural fluid and her bronchial passage. She had anthrax. Although deemed well enough to leave the hospital, for Jyotsna and her family the misery continued. Nine months later Ramesh said, "We have been through hell. You have no idea." He recounted the continuing toll:

> In the middle of the night she wakes up screaming: "They're coming to kill me." Her mind is gone. Memory is a big problem and she is confused a lot. But the biggest problem is fatigue. She cannot walk to the end of the block without resting. She was a workaholic type, and now she doesn't feel like doing anything. And she is afraid to touch the mail. Even now, she's not touching the mail.

Ramesh reflected on the effects on Plita, at 19 the older of their two daughters. A sophomore at New York University, she can hardly bear to call home. Whenever she does, Ramesh said, "She would cry over her mother for an hour." In the midst of the ordeal Plita developed colitis, all while trying to attend to her studies. "For us," Ramesh says, "like I said before, it has been hell."

The Hamilton postal distribution center, 20 miles north of Willingboro and 5 miles south of Princeton Junction, had sunk the lives of Norma Wallace and Jyotsna Patel into a dismal morass. Along with the center on Brentwood Road in Washington, D.C., it would prompt Americans to think as never before about their mail and the system that delivered it.

Months after the closing of the Brentwood mail processing center in northeast Washington, a security guard warded off a curious passerby, me, from approaching its front gate. "And no pictures," he added after spotting the camera slung from my shoulder. Farther down the street, far from the guarded entry, one side of the single-story building can be approached more closely. The long red brick wall is interrupted by only a metallic sheeting that covers every portal—windows, doors, delivery platforms. The rigid sheeting was placed in anticipation of decontamination efforts, then scheduled to begin in mid-2002.

Apart from its size, the most striking feature about the Brentwood facility was the sense of normal activity nearby. Unlike the Hamilton center, which is on a highway, the Brentwood building is in the middle of a residential area. Hundred-year-old attached homes line Rhode Island Avenue, a couple of blocks from the facility. Directly across the street, on Brentwood Road, a gas station stands on a corner adjacent to a row of stores: a delicatessen, a Social Security Administration office, the Happy Face Child Care Center, a dry cleaners. Fifty feet from the stores, a McDonald's restaurant bustled with patrons. Like the people moving along the street, the diners appeared oblivious to the anthrax-contaminated colossus across the way.

Inside the Brentwood building, as in Hamilton, dozens of machines stood idle. Leroy Richmond thought about the machines, the mail that went through them, and how he might have become infected. Every morning at 3:00, before he became sick, he would begin his commute north on Interstate 95 from his home in Stafford, Virginia. An hour later, at Brentwood, he would board the mail truck for a 50-minute ride to Baltimore-Washington International Airport. There, he and eight co-workers would sort by hand express mail that had been flown in and then take it back to Brentwood. The job was completed by 10 a.m., after which he could help elsewhere as needed.

On Thursday morning, October 18, following his truck run to the airport, Richmond's supervisor called him over. "Rich, I'd like you to help straighten up before Postmaster General Potter gets here." Actually Richmond had been doing just that since the previous week—organizing boxes and bins and cleaning various areas of the building. In the wake of the discovery of the Daschle letter, Potter would be arriving that afternoon to assure the workers that they were not at risk. Although Richmond was feeling sluggish, he continued to work, never imagining his illness might be connected to that letter.

The anthrax in the letter sent to Senator Daschle was found to have been of an unusually fine grade. When the envelope was opened in the senator's office on October 15, the powder floated out as if lighter than air. The material seemed to be a finer mix than in the previously located anthrax letters. Like the two earlier letters that had been recovered—those sent to Tom Brokaw and the *New York Post*—the Daschle letter bore a postmark from Trenton. But the postmarks on the New York-bound letters were dated September 18. The one on the letter to Daschle read October 9. An anthrax letter to Senator Patrick Leahy, also postmarked October 9, would later be found. From the Hamilton facility, the Daschle and Leahy letters were moved to Washington and processed at Brentwood on Thursday, October 11. "That Friday and that Saturday I was assigned to do some work behind the sorting machines," Richmond recalled.

> The mechanics were blowing the machines out with air pressure. I was working less than 2 feet from the machine that ran the Daschle letter. Possibly that's when I was exposed. Actually I worked there again in the course of the week, so I could have been exposed repeatedly. The machine was so contaminated with spores.

How might have the other three men been infected? Leroy Richmond had thought about this before. No one knows the answer with certainty, but his response was as informed and plausible as any. As for Qieth McQue's exposure:

> They quarantined the mail right outside of his office, leading to the workroom area. That's where they quarantine all the government mail. So it's a possibility he could have gotten infected there. Or

> maybe on the workroom floor, in the vicinity of where I was working. He had to pass that area.

What about Joseph Curseen and Thomas Morris, the men who died?

> They were different. Joe Curseen ran the Daschle and the Leahy letters on his machine. How do I know that? Because when the letters go through the machine there is a marker left on the letter that tells you the machine that ran the letter. Since it was mail going to a government agency, it is separated and worked on by a particular machine. And there is a computerized printout that identifies all the letters that go through a machine. The mere process of loading the machine was how he was probably infected.

And Mr. Morris?

> In his case, he would have to verify all the mail going to the Senate office buildings and put it into a particular tray. You know—this mail is for Senator Daschle, this is for Leahy, for Senator Rockefeller, whomever. He would take the mail in his hands, maybe 20 letters at a time, and ruffle through them. That's how he probably got infected.

The source of David Hose's exposure is also unproven, though it almost certainly came from the State Department mail annex in Sterling, Virginia, where he worked. By the time Hose walked into the Winchester Hospital emergency room on Wednesday, October 24, he was convinced he had anthrax. Reports about the four infected Brentwood workers, including the two who had just died, were filling the airwaves. Hose recalled that in recent weeks, "we had handed over at least six suspicious letters to the sheriff's department to deliver to the FBI to check out." After the discovery of the Daschle letter, mail in several government agencies was set aside and eventually filled 280 55-gallon drums. On November 16, as investigators in respirator masks and protective suits were culling through the batches, they found an envelope that was addressed in handprinted capital letters to Senator Patrick Leahy. Like the Daschle letter, it was postmarked October 9, Trenton, New Jersey. The address was also in block letters: 433 Russell Senate Office Building, Washington, D.C. 20510-4502. The bar-code machine had mistakenly read the fourth digit of the zip code as "2" instead of "1," which resulted in misrouting the letter to the State Department. During the next 3 weeks, the letter was analyzed with exquisite care. At the laboratory of the U.S. Army Medical Research Institute of Infections Diseases in Fort Detrick, Maryland, a small hole was bored through the envelope and the powder drawn out by suction. On December 6 the army scientists confirmed that the envelope contained powdered anthrax of the type that was in the Daschle letter. The message sheet inside was also identical to the one sent to Daschle.

Months later, Hose reflected on what might have happened in his mailroom at the State Department annex:

> The letters went through sorters over at Brentwood, and we got all our registered and regular mail from Brentwood. I was in charge of the incoming mail unit here at the annex and supervised six people who would open up the pouches of mail. I would sometimes handle the mail myself. I feel sure that Leahy's letter was one of the six letters I handled and that we gave to the sheriff's department.

The precise moment that the anthrax victims became infected will always remain uncertain. The likely time for the Brentwood and State Department inhalation victims can be narrowed only to a period of days. But the moment of exposure for the two New Jersey inhalation victims may have been when Norma Wallace believes it was, at 5:30 p.m. on Tuesday, October 9.

Norma was on the late shift at the Hamilton, New Jersey, postal center, from 3:30 p.m. to midnight. Two people usually operate the machine she was working on, but that day she was alone because "we were short on personnel." Every 2 hours the machine operators get a 15-minute break. Shortly before her first break at 5:30, the machine jammed. "The electronic eye was covered with dust," she recalled, "and the machine stopped. Paper dust accumulates if there is a high volume of mail and if the machine is not cleaned after the previous shift." She notified a mechanic, who came over to blow the dust out of the machine. "There are coiled air hoses suspended from the ceiling," she said. "He just pulls one down and blows the machine with compressed air."

During the seconds the mechanic was blowing the air, Norma stood a few feet away with her back to the machine. At the same time, Jyotsna Patel approached her. "Hi, Norma. I'm ready to relieve you," she said, knowing that Norma was about to take her break. "Jyotsna probably didn't think about walking through a cloud of dust," Norma recalled. "We both stood there and I told her what I was up to and where the mail was and for her to take over."

Much of this story comes from Norma. Jyotsna's memory loss is more severe, and she is also less comfortable about discussing the incident.

Joseph Sautello, the acting Postmaster of the Hamilton center, said that the precise time and location that Norma and Jyotsna were exposed remain uncertain. But he did confirm that computer records indicated the Leahy letter passed through a sorting machine at 4:57 p.m. and the Daschle letter at 5:15 p.m.

The total number of anthrax cases in New Jersey amounted to six— two inhalation and four cutaneous. Five of the six were postal employees, four of whom worked at the Hamilton center; the other was a letter carrier from the West Trenton post office. The sixth infected person, Linda Burch, was a bookkeeper at a Hamilton Township accounting firm that received mail directly from the postal center. None of the patients died, though a year later, among the five postal workers, Teresa Heller was the only one who felt well enough to return to work.

On Sunday, October 21, three days after the Hamilton center was closed for testing, the New Jersey Department of Health and Senior Services issued a press release. It included a stunning announcement by Dr. Eddy Bresnitz, the state epidemiologist: "At the Hamilton Township facility, 13 out of the 23 samples collected by the FBI tested preliminary [sic] positive for anthrax." The worries expressed at the October 18 meeting by George DiFerdinando and others that the building might be contaminated indeed proved valid. Additional sampling in the following months by the state health department, the U.S. Environmental Protection Agency, and the CDC showed that anthrax was rampant throughout the facility. Moreover, testing of the 50 local post offices that fed into the Hamilton center later confirmed the presence of anthrax in five. But the early finding alone was unnerving and entirely unexpected.

For 20 years, during the 1950s and 1960s, the U.S. Army conducted hundreds of germ warfare tests in populated areas throughout the United States. Mock biowarfare agents were released from boats, slow-flying airplanes, automobiles, germ-packed light-bulbs, perforated suitcases, and wind-generating machines. The test agents included the bacteria *Serratia marcescens* and *Bacillus subtilis*, and the chemical zinc cadmium sulfide. (Although less dangerous than real warfare agents, the test bacteria and chemicals posed risks in their own right.) Cities and states were blanketed, including San Francisco, Minneapolis, St. Louis, and parts of Illinois, Ohio, and

Hawaii. Some attacks were more focused, such as those in which bacteria were released in the New York subway and on the Pennsylvania Turnpike. In each instance the spread and survivability of the bacteria were measured to assess the country's vulnerability to a biological attack. Apparently the testers never considered the U.S. mail a possible vehicle.

A 1999 study by a defense-contractor did involve opening an envelope containing 2.5 grams of *Bacillus subtilis* in an office. Details are not available, but the test evidently did not foretell the massive cross-contamination that would occur with the anthrax letters. (The investigation was conducted by William Patrick, who, before the United States ended its biological warfare program in 1969, had helped develop the American germ arsenal. The 1999 study had been commissioned by Steven Hatfill, who was then an employee of the San Diego-based Scientific Applications International Corporation. Hatfill later became a much-publicized subject of the FBI's search for the anthrax mailer.)

In October 2001 it was becoming clear that the mail might be a far more simple, inexpensive, and efficient means of germ dispersion than any of the previously tested methods. The most stunning realization was that spores could escape from tightly sealed letters, a fact that no one before seems to have contemplated. A 1- to 3-micron-sized anthrax particle could pass through a paper envelope, the pores of which can be 20 microns in size. Air currents could then whip the leaked spores in any number of directions. In a postal center, spores might settle not only in the nasal passages or on the skin of workers but also on sorting machines and other mail. Each newly infected piece of mail could in turn become a potential carrier to other mail and other people.

The transmissibility of the spores prompted health and law enforcement authorities to use a new term that gained quick and frightening currency: cross-contamination. From the time that the first anthrax letters were processed on September 18 through the time that the Hamilton and Brentwood buildings were closed, 85 million pieces of mail were handled at the two facilities. They were bound for destinations in every state of the union and in countries around the world. No one knows how many letters became cross-contaminated, but it is likely that thousands, perhaps millions, might have carried some anthrax spores. Yet few people showed evidence of infection. And in the known cases, at least the source of infection could be traced to infected mail. Stephen Ostroff, chief epidemiologist for the CDC, described the feeling among his associates: "I think in the back of everybody's mind was the question, 'What's the potential for massive risk?' Still, it was comforting that all the exposures were explainable."

But if massive infections never materialized, the comfort about explainable exposures was short lived. In due course, two more anthrax cases were identified, each without a clue about how the victim was exposed.

CHAPTER
FIVE

THE OUTLIERS

The sound of salsa music floats into the lobby from a ground-floor apartment at 1031 Freeman Street in the South Bronx. The 30 brass mailboxes in the entranceway largely bear Spanish-sounding names like "Espina," "Rivera," and "Rodriguez." The name on the box for Apartment 3R, "Nguyen Xinh K," is an exception, not only because the name is Asian but because its bearer is no longer alive. A month after Kathy T. Nguyen's difficult death from inhalation anthrax on October 31, 2001, her Vietnamese name remained on her mailbox.

Kathy's mailbox was interesting for another reason as well. Investigators had found no trace of spores in her clothing, apartment, workplace, or mailbox. Still, a nagging suspicion remained that her death was somehow connected to the mail. After all, until then every instance of anthrax infection since the 4-week-old outbreak began had been traceable to mail.

Above the row of mailboxes was a picture of Kathy, faintly smiling. It was posted beside a request by the FBI, the police, and the New York City Department of Health: "Anyone who saw her or knew of her movements and activities between October 11, 2001 and October 25, 2001 please call the Task Force Hotline: 646-259-8539." In the weeks after Kathy's death, the hunt for the source of her infection became intense.

Kathy had arrived in the United States in 1977, a refugee from the collapsed South Vietnamese regime. Since 1990 she had worked at the Manhattan Eye, Ear, and Throat Hospital on the Upper East Side, where she delivered supplies from the stockroom to the clinics and operating rooms. Adored by neighbors and co-workers, this small unassuming woman could not know that at age 61 her last desperate breaths would define her tragic celebrity.

"Dave, I'm not feeling well. It hurts to breathe," Kathy said late Sunday morning, October 28. Dave Cruz had managed the Freeman Street apartment complex since 1981 when Kathy moved there. Her words were still fresh in his mind weeks after her death. From his storefront office around the corner from the apartments, Dave shared his thoughts and his sadness. A husky figure wearing a T-shirt, he has a surprisingly soft voice. Kathy lived alone. She was undemanding, he said, paid her rent on time, kept her apartment spotless. "You couldn't get any better than that."

Dave was already on his way to Kathy's apartment that Sunday morning when they encountered each other in the courtyard adjacent to the building. One of his workers had found him to say that Kathy was urgently looking for him. When he saw her he asked, "Would you like to go to the hospital, Kathy?" thinking of Jacoby Hospital a half mile away. "Yes, but please take me into Manhattan to Lenox Hill." Dave rushed for his minivan.

Lenox Hill Hospital is on East 77th Street, 13 blocks north of its smaller subsidiary, "Eye and Ear," where Kathy worked and where elective surgery was the principal fare. Kathy's own physician was on the staff of both hospitals. After a decade of commuting, she knew the area well. Her trip to work ordinarily began a block from her house at the Whitlock Avenue subway station. Twenty-five minutes on the Number 6 train to the East 68th Street Station put her just 5 minutes by foot from work.

But that Sunday, Dave reduced her usual commute time by a third: through some traffic lights along Bruckner Boulevard, over the Third Avenue Bridge to Manhattan, and a final uninterrupted stretch along the FDR Drive. As he drove, barely aware of the sun's reflection off the East River, Dave remained focused on Kathy. He was incredulous when she tried to hand him money orders for 2 months' rent—in case of a long hospital stay, she said. "Put them back in your purse, and don't worry about such things now." Kathy's labored breathing prompted questions. Did she have asthma? Bronchitis? he asked. "No, no, it just hurts." He tried to reassure her: "Don't worry. Maybe they'll give you antibiotics." At 11 a.m. they arrived at the emergency room, and he walked her into the examination area. Dave Cruz did not know what else to do except give her his business card and a word of assurance before leaving. "Kathy, please call me if you want anything."

Dr. Mayer Grosser spends much of his time in a short-sleeve green shirt, the pajamalike top of standard hospital garb. A 13-year veteran of Lenox Hill's emergency room, he is used to seeing people in distress. Kathy's case did not seem extraordinary. She described her symptoms—chest pain and shortness of breath that had worsened during the previous 48 hours. He put his stethoscope to her chest and back and heard crackling sounds across her lungs, an indication of fluids. "Rales," he thought to himself. A chest X ray also suggested pulmonary edema, the presence of fluids in and around the lungs, all of which he thought pointed to the probability of congestive heart failure.

Kathy worried that her illness and her absence from work would cause some inconvenience to her employer. "How can we get in touch with my boss?" she wondered. "I have to tell him I'm not going to work tomorrow." "Don't worry," Dr. Grosser assured her. "We'll take care of that." Meanwhile, he pondered over her medical history. The only hint of abnormality was her elevated blood pressure, but that was under control with daily pills. After a few hours on nitrates and a diuretic for presumed heart failure, there was no improvement. An echocardiogram later confirmed that she had no heart abnormalities.

About 5 p.m. Grosser looked again at her chest X ray. "I saw the widened mediastinum and it hit me. I thought, 'This could be anthrax.'" He asked her if she worked near the mailroom. "Yes," she said and explained that the mailroom was attached to the central supply room. "I never mentioned the word 'anthrax,' and she didn't either," he said. "But she was a smart woman and she became visibly concerned after that." By then he had her taking large doses of Levaquin, an antibiotic in the ciprofloxacin family. At 7 p.m. Kathy Nguyen was moved to intensive care.

By the time Dr. Bushra Mina saw her at 9 p.m., Kathy's breathing was more difficult. She was lucid but told him she was very tired. Mina, a pulmonary and critical care specialist, was aware of Grosser's presumptive diagnosis. Slightly built and soft spoken, Dr. Mina recalled how worried he was about Kathy's unusually rapid deterioration. "She seemed to be weakening by the minute," he said, and he ordered an intubation tube to help her breathe. Inserted through her nose into her trachea, the tube also prevented her from speaking. Around midnight she seemed stable, and Mina went home for a few hours of sleep.

When Dr. Mina returned at 6 a.m., he had Kathy sedated while he performed a bronchoscopy, which enabled him to peer into the small passageways that branched throughout her lungs. His viewfinder became immersed in bloody, infected fluids. During the remainder of the day, despite placement of tubes to drain the fluids from her chest, her condition worsened. She lapsed into unconsciousness, and her organs began to fail.

Meanwhile, on that same day, Monday, the results of testing at the large Morgan mail distribution center at 9th Avenue and 29th Street in Manhattan were announced. Five of the sorting machines had small amounts of anthrax, though no postal workers had become infected. The letters that had caused the seven cases of anthrax in the New York City media offices apparently had been routed through those machines. The two recovered letters, to NBC and the *New York Post*, had first been processed on September 18 at the Hamilton center in New Jersey and then forwarded to Morgan.

Now, the news about Kathy's inhalation infection unnerved everyone. Could she have been exposed to anthrax from cross-contaminated mail that, 40 days earlier, had gone through Morgan? Was a new attack under way? The New York Metro Area Postal Union urged that the Morgan facility be shut down. But the postal authorities closed off only the contaminated sections on the second and third floors. David Solomon, a postal service spokesman, said the building would remain open because "health professionals [have not said] there is a health danger for our employees." Still, workers were worried. The absentee rate at the facility climbed to 30 percent.

As word circulated in both hospitals about Kathy's anthrax, anxiety among hospital personnel also rose. At the request of the city health department, patients from Manhattan Eye and Ear were moved out and the hospital was closed. A week later, after no anthrax had been found there, the hospital reopened. During the closing, 250 surgeries were canceled, and by the end of the week the hospital had lost about $1 million in revenue. Dr. Thomas Argyros, the medical director at Lenox Hill, recalled the atmosphere at the time:

> The dynamics were terrifying because of the possibility that she got the anthrax at the hospital. Anyone who had been in Eye and Ear since around October 15 had to go for prophylaxis. About 2,000 people were screened here [at Lenox Hill] and at [nearby] New York Hospital, and they went on antibiotics.

One of the people screened at Lenox Hill, Daisy Cruz, had been to the Eye and Ear Hospital on October 18 for a hearing test. At her screening for anthrax, she said that she was "crazy" with fear: "I see anthrax in the toothpaste. I see it in the orange juice. I see it in the sugar. They're going to kill me with a heart attack before they kill me with anthrax."

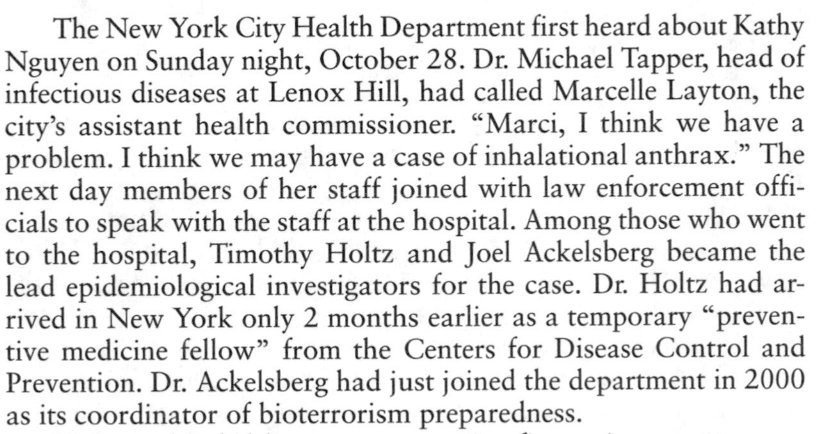

The New York City Health Department first heard about Kathy Nguyen on Sunday night, October 28. Dr. Michael Tapper, head of infectious diseases at Lenox Hill, had called Marcelle Layton, the city's assistant health commissioner. "Marci, I think we have a problem. I think we may have a case of inhalational anthrax." The next day members of her staff joined with law enforcement officials to speak with the staff at the hospital. Among those who went to the hospital, Timothy Holtz and Joel Ackelsberg became the lead epidemiological investigators for the case. Dr. Holtz had arrived in New York only 2 months earlier as a temporary "preventive medicine fellow" from the Centers for Disease Control and Prevention. Dr. Ackelsberg had just joined the department in 2000 as its coordinator of bioterrorism preparedness.

Holtz and Ackelsberg are contrasting figures in appearance as well as background. Holtz's closely cropped, fire-red hair, mustache, and beard seem out of phase with his laid-back demeanor. His baritone voice is flat and even, revealing few signs of emotion. In contrast, Ackelsberg, in blue jeans and rumpled green jacket, necktie askew, speaks in sonorous tones, a hint of the actor he once aspired to be.

Their paths to medicine could hardly have been more dissimilar. Holtz's was all Iowa. He grew up there and went to college and medical school at the University of Iowa. For Ackelsberg, at age 50—12 years older than Holtz—medical training came much later. "I had been a typical 60s kid who had gone to a number of schools in a number of places," he said. Before completing Tufts Medical School in 1993, he had variously been an actor, a carpenter, and for 10 years in San Francisco a cab driver. Both men went on to train in public health, and both served for 2 years with the CDC's Epidemic Intelligence Service before coming to New York City. Now, the two would work in tandem, medical sleuths in search of the source of Kathy Nguyen's anthrax.

On Monday morning, when Holtz first heard about the suspicion that there was a new inhalation anthrax case, he thought, "Nah, that can't be possible." Like others in the department, he thought New York's seven cutaneous cases were all they would be seeing. Later that day, the department learned that the new case had been confirmed. Holtz recalled his feelings:

> Marci Layton asked me to get involved and be sort of the head of the epidemiology investigation at the patient's workplace, and I thought, "Okay, here we go again." It's like it's not over. It was one thing to have cutaneous cases, but when you have an inhalation case, well we were all extremely worried about what was going to happen that week, that this was a harbinger of worse things to come.

Under Layton's overall coordination, Ackelsberg and Holtz became "sort of the two team leaders," in Holtz's words. After their first visit together to Lenox Hill they divided responsibilities. Ackelsberg: "I was focusing more on interviewing friends, seeing her apartment, trying to learn about her and her activities." Holtz: "I was the one who mainly spoke to her co-workers, people at the hospital she worked at."

Holtz recalled that when investigators met to interview workers at the Eye and Ear Hospital, agents from the Federal Bureau of Investigation were reluctant to enter the building. They were concerned that it was contaminated, though the health department staff felt the risk was minimal. "In the end, there were two FBI agents who agreed to go inside with us. The rest of them stayed across the street in a hotel and did some interviews over there," said Holtz.

One of the people that Ackelsberg interviewed was Anna Rodriguez. Weeks after Kathy's death, Anna was still stunned. She had known Kathy since the early 1980s when Kathy moved to the newly renovated Bronx apartment building. "A classy lady," Anna said about Kathy, "always neatly dressed." And her apartment? "It was so clean you could eat off the floor." Kathy's monthly rent, which she paid by postal money order, was $675. The building was a Section 8 apartment house, which meant that her rent was 30 percent of her gross income. ("Section 8" refers to a federal program that provides housing subsidies for low-income families.) Kathy's annual income was about $27,000.

"I used to help her with filling out applications and money orders," Anna said. Kathy was scrupulous about meeting her obligations, and her rent was always paid on time. Her heavily accented English could be difficult to understand until you got used to it, Anna said. But she conveyed a kindness that made people feel comfortable.

Anna and Kathy would come across each other on the elevator or in the triangular courtyard outside the building's entrance. "Every time I saw her she would ask how I was, how my family was." She and Kathy had been to each other's apartments over the years. "Kathy once made wonton soup and fried rice for me," said Anna. And Kathy often gave little gifts to the neighbors—fruit, a handkerchief. But even after all those years Anna wasn't sure how well she really knew Kathy. "Everyone liked Kathy," Anna said, "but you never saw her with anyone else. She was always alone." Kathy seemed to be entirely on her own, "no relatives that I knew, though I heard she had a cousin in Seattle."

Anna Rodgriguez lives in 5R, two floors above Kathy's apartment. Her modest-sized living room is lively with color, family pictures, and cabinet shelves lined with decorative dishes and vases. "The room layout in Kathy's apartment is the same as mine."

Three weeks after Kathy's death, across the corridor from the third-floor elevator entrance, the heavy black door to Kathy's apartment remained bolted. Her picture and a brief article about her, clipped from a Spanish-language newspaper, were taped to the door. A separate notice said: "These Premises Have Been Sealed by the New York City Police Department." Beneath the notice a peephole allowed for a partial view of the interior. The livingroom rug was gone, and the floor's wooden slats were bare. Furniture had been moved from the center area. Against the wall on the left side, a 3-foot-tall shelf supported a plant whose leaves were withered beyond recovery. Farther along the shelf were a coffee mug, a small basket, and a few books. Toward the far wall a low wooden stool was visible.

Investigators who visited Kathy's apartment after she was diagnosed with anthrax were impressed by how much was in it. Joel Ackelsberg was part of the team looking for clues. Months later he recalled his impressions: "There was a lot there. She saved mail and old receipts." Trinkets were plentiful, also lots of clothes and shoes. "I saw two dozen shoeboxes," he said. The CDC's Stephen Ostroff described Kathy as a "conscientious saver." Ostroff saw neat piles of "junk mail, some from 10 years ago, that were asking for donations." Since she kept the apartment extremely clean, Ostroff wondered if she might have wiped out any anthrax that might have been there.

The fact that Kathy's anthrax at first seemed unconnected to the mail was, to some, strangely encouraging. By tracing where she had been in recent weeks—shops, parks, subway lines—investigators thought the trail might lead to the perpetrator. They believed that "along those routes she might have intersected with the person or people behind the anthrax attacks," according to a *New York Times* report. No less intriguing was her place of work. The basement stockroom at the hospital was next to the mailroom. But a few weeks after her death, none of the hundreds of tests for anthrax there or elsewhere showed any indication of anthrax. Nor were the interviews with friends and co-workers any more helpful.

Law enforcement officials had even tracked down a former husband. Kathy had been married briefly some years earlier, though she had not seen her ex-husband or other relatives for decades. Eager for leads, FBI agents managed to locate him. He told them that Kathy used to like to smell things. "We certainly were intrigued by that," said Ackelsberg. The team considered flowers and anything else she might have smelled as a possible source of the anthrax. "But, again, nothing came up."

Kathy's decade at the Eye and Ear Hospital had left an impression similar to the one held by her neighbor, Anna Rodriguez. Esperanza Vassello, the hospital's head dietician, had been at the hospital since emigrating from the Philippines 30 years earlier. A few weeks after Kathy's death, "Espie" sat in her office, next to the large kitchen she supervises. She had a strong face and spoke with conviction. "See that?" she asked me, pointing to a brown-framed desk clock on a shelf to her right. "Kathy gave it to me for my birthday." She recalled the conversation she had with Kathy less than 2 months earlier.

"Espie, here is your birthday present," Kathy said.

"But it's only October 8 and my birthday is November 7."

Kathy smiled. "I want you to use it longer. Happy birthday."

Kathy gave presents, and not just for birthdays, to everyone at work—earrings, cologne, food. Espie noted that Kathy "had a problem with language and understanding forms." So, as she did with Anna Rodriguez, Kathy would ask Espie for help in filling out money order requests and other forms. Espie felt that they had become closer during the previous year or so. But she was surprised to learn from a newspaper article after Kathy's death that she had a cousin in Seattle. "She had told me that all her family members were killed in the Vietnam War and that she had no relatives here."

Ackelsberg summarizes the view of the investigative teams:

> The impression we had was that she was a woman who worked and lived according to a set schedule, that she missed very few days of work. She was very friendly. She was considered extremely generous. At the same time, the people we were directed to as being really close friends—well, if they were as close as anyone got to her, then she lived a fairly isolated existence.

Was Kathy Nguyen's end foreordained when she swung open the little brass door on her mailbox and retrieved a letter dotted with particles of death? Or did she inhale a lethal dose from some other source?

By the time of her death, 4 weeks had elapsed since Bob Stevens had become the first fatality of the anthrax terrorism. During those weeks, eight more people became infected with inhalation anthrax and 11 with the less dangerous cutaneous form. Thousands of people were hoarding Cipro and other antibiotics deemed effective against anthrax. Gas masks were selling briskly, even chemical protective suits. People were opening mail in their garages and scrubbing their mailboxes with bleach. Contaminated postal facilities, business offices, and the Hart Building, which housed the offices of half the U.S. Senate, were closed. Postal authorities reported a sharp decline in the volume of mail.

CDC protocols state that health care workers caring for anthrax patients are not at risk and that they need not take antibiotics. Nevertheless, after Kathy Nguyen was diagnosed with anthrax, both Dr. Mayer Grosser and Dr. Bushra Mina, who had seen her at the hospital, started on Cipro. After a few days, Grosser was inclined to stop, but his wife would not let him. "She insisted on watching me take the Cipro, one pill in the morning and one at night."

On Wednesday, October 31, Kathy Nguyen went into cardiac arrest and at 1 p.m. she died. Blood cultures and autopsy findings confirmed that anthrax had colonized all her organs. On the day of her death, the *New York Times* ran stories about Kathy Nguyen under a three-column headline: "Hospital Worker's Illness Suggests Widening Threat; Security Tightens Over U.S."

A week after Kathy's death, no new cases had been reported. "I'm hopeful, like the rest of America, that the anthrax has stopped permanently," said Tom Ridge, President Bush's director of homeland security. But by mid-November, hopes that the outbreak had ended were abruptly set back.

Ten miles south of Waterbury, Connecticut, Route 67 leads to a small wooden bridge and a sign that says "Welcome to Oxford." Oxford was established in 1789, the year George Washington was inaugurated president of the United States. Two centuries later the number of inhabitants had grown from 1,400 to 9,800, a comparatively modest increase. During the same period in New York City, then as now the nation's largest city, the population rose from 33,000 to 8 million. Located 75 miles north of New York, Oxford is a rustic community where agriculture and farming continue to flourish.

Along Great Hill Road, just past Immanuel Lutheran Church, a road to the right opens to a patch of single-family homes. One such home, on Edgewood Drive, stands behind neatly trimmed hedges and a tall flagpole. A black mailbox bearing the number 16 is perched on a white post near the curb. Toward the rear of the house, the backyard slopes into a valley of maple and oak, naked in the January cold. The house is empty. Six weeks earlier, on November 21, 2001, its owner, Ottilie Lundgren, succumbed to anthrax. At 94 she was the oldest of the 11 inhalation anthrax cases. Although the source of her infection, like that of Kathy Nguyen's, remains uncertain, Ottilie's seems even more puzzling. Her activities were limited, easily traced, and seemingly unrelated to any source of anthrax.

Whereas Kathy Nguyen had been part of the hubbub of New York City—the crowds, the subway, the shops in Chinatown—Ottilie largely remained at home in a rural, quiet community. "We were left with major gaps in the time line of Kathy's activities," Joel Ackelsberg said. This was not the case with Ottilie. She was childless and since her husband's death 24 years earlier, she lived alone. Neighbors would stop by, as did her niece, Shirley Davis, who lived 30 minutes away in Waterbury, Connecticut. Ottilie had given up driving a year earlier and had become dependent on them to help her with shopping. They would drive her to the church down the road, to the beauty parlor, and to lunch at Fritz's Snack Bar on Highway 67 where she would order minestrone soup and lobster rolls.

Although slowed by age, Ottilie remained remarkably alert. At one time a manager of her husband's legal office, she was still a quick thinker and a reader of travel and mystery stories. Robin Shaw, a half century younger, had for the past 10 years been Ottilie's hair dresser at the Nu Look Hair Salon. "She was a sweet woman, intelligent, and very frank," said Robin, who looked forward to seeing her every Saturday morning at 11. Ottilie could effortlessly shift from reciting poetry to gentle chiding. A fastidious, conservative dresser herself, she may have seen in Robin's fluffy dark hair and colorful clothes some opportunity for improvement: "Robin, your outfit. Couldn't you choose a better color match?" "Now don't hold back, Ottilie," Robin replied. And both broke into laughter.

Weeks after Ottilie died, Shirley Davis talked with me about her aunt over lunch. Dressed in a black suede jacket and maroon sweater, a thin gold necklace below her turtleneck collar, Shirley's meticulous appearance is reminiscent of her aunt's. At 72, Shirley, who had reared her own children, was young enough to have been Ottilie's daughter. "We became especially close in recent years," Shirley says. Born and raised in Waterbury, like Ottilie, Shirley still lives there. She visited Ottilie several times a week to check on her and help around the house. Only after her death did a ritual of Ottilie's take on significance:

> Someone would bring in the mail from the mailbox along the curb. And I watched many times as Ottilie sorted it. Sometimes I sat at the card table with her. She would open each letter with a letter opener. Every letter that she did not save, she would tear in half, into two pieces, and discard it into a wastebasket underneath the card table. Sometimes I would rip them for her. She filed and organized all the mail that she kept. Then she would empty the wastebasket into a garbage bin outside the back door.

"I nearly fainted when the doctors told me they suspected anthrax," Shirley said, her eyes beginning to water. How could her aunt ever have been exposed? Her life was distant from mailing facilities and population centers. Her infrequent outings never took her more than a few miles from home. Shirley recalled that Ottilie had been feeling out of sorts for a few days—mild cough, weakness—and that she rejected Shirley's pleas to call her doctor. But with Shirley in her home on Friday morning, November 16, she relented. Ottilie's doctor suggested that she stop by the emergency room at Griffin Hospital in nearby Derby.

William Powanda, the burly gray-haired vice president at Griffin, recounted Ottilie's relatively benign symptoms: low-grade fever, slight dehydration. "Classically, she would not have been admitted, but because of her age we decided to let her stay." The remainder of that day and into the next Lundgren barely seemed ill. She joked with visitors and medical attendants; her initial tests suggested a possible urinary tract infection. On Saturday morning Dr. Lydia Barakat, one of the hospital's infectious disease specialists, received a call from the intern on duty. Lundgren's overnight blood cultures had grown bacteria that were rodlike and tested Gram positive. Barakat had not yet seen Lundgren, but she thought "elderly patient; probably clostridium." She ordered Ottilie on antibiotics, vancomycin and ceftazidime, as she set off to the hospital to see the patient. Later she added oral Cipro and intravenous ampicillin to the treatment.

Clostridium bacteria, at least the type most often found in humans, are rarely life threatening. When Barakat arrived at the hospital around 10 a.m., she went to the laboratory, checked Lundgren's reports, and looked at her chest X ray, which appeared normal. She peered through the microscope and saw the cultured bacteria—all consistent with clostridium. Barakat, an attractive young woman who attended medical school in her native Lebanon, had completed her residency at Griffin only 2 years earlier. That Saturday morning, the case hardly seemed extraordinary, certainly nothing to warrant interrupting the weekend of the hospital's infectious disease chief, Howard Quentzel. Still, in the back of Barakat's mind lay another truth: Gram- positive rods are also consistent with anthrax.

"I was thinking clostridium, but I had been hearing a lot about anthrax," Dr. Barakat later said. She went to Ottilie's room. While examining her, Barakat asked, "Did you receive any mail with powder in it?" Ottilie answered, "No." She did not seem very ill and had no respiratory problems, and Barakat was reluctant to say "anthrax" to anyone else. "I did not want to make a fool of myself," she told me. On Sunday Lundgren seemed more short of breath. A new chest X ray, unlike the earlier one, suggested that fluid might be present in one lung. By Monday morning, after more tests, it became clear that Ottilie's infection was not from clostridium but from a bacillus.

The new tests showed that the bacteria were nonmotile and nonhemolytic—that is, they did not move spontaneously and did not destroy red blood cells. This ruled out *Bacillus cereus*, a treatable and rather common infective agent. Meanwhile, Barakat had briefed Dr. Quentzel about the patient, and he was eager to see her when he arrived at the hospital on Monday. Another chest X ray had just been taken, the third since Lundgren entered the hospital. This one showed fluid throughout both lungs, though no widened mediastinum.

Howard Quentzel, at Griffin for 13 years, had graduated from New York Medical College and completed his residency in infectious diseases at Lenox Hill Hospital. There he had worked under Dr. Michael Tapper, the man who, weeks earlier, had overseen the care of Kathy Nguyen. Quentzel's slight build, high forehead, and large glasses present a scholarly appearance. Earlier he had reviewed his textbook to make sure which bacilli are nonmotile and nonhemolytic. The prime candidate: *Bacillus anthracis*. He and Barakat met to review Lundgren's chart. He turned to her and said: "This is anthrax."

"I'm relieved," Barakat thought to herself. Barakat felt as though a burden had been lifted from her. Now someone else also accepted the seemingly preposterous notion that Ottilie Lundgren, this stay-at-home nonagenarian, could have anthrax. But what else could it be? Quentzel went to see Ottilie, who was still quite coherent. "Did you travel to New York?" he asked.

"No, I haven't," Ottilie said.

As did Barakat earlier, he asked, "Did you see any powder in the mail?"

"No."

Barakat and Quentzel understood the explosive implications of their diagnosis. The country was already in turmoil over anthrax. The Lundgren case would add a new dimension to the fright. If someone so unlikely could become a victim, everyone might be at risk. If Ottilie's illness were connected to a letter contaminated by an already cross-contaminated letter, the implications were profound indeed. Her letter then could have been part of an infective chain reaction. Third- or fourth-generation cross-contamination could have devastating implications everywhere. Even the limited comfort in knowing that anthrax did not spread from person to person could be undermined. Routine mail might become the anthrax equivalent of the contagious smallpox cough or the touch of a plague-infected hand.

An ominous comment in 1981 by Rex Watson, then head of Britain's chemical and biological defense establishment, seemed almost an understatement. If Berlin had been bombarded with anthrax in 1945, he said, the city would still be uninhabitable. But he was speaking of massive quantities of spores. Now the United States seemed to be hostage to a few grams. The randomness, the far-flung locations, and the uncertainty of who might be next overshadowed the limited number of victims thus far.

Barakat and Quentzel knew their lives were about to be disrupted, as were those of others in the hospital, not to mention Ottilie Lundgren's neighbors, friends, and family. They decided on a division of responsibility. Barakat would focus on taking care of her patient and try to disregard the inevitable distractions. Quentzel would notify the appropriate hospital authorities, state health officials, and the federal government through the CDC. "I was excited," Quentzel acknowledged in reflecting on that moment. "I knew what this meant."

Quentzel's calls to state and federal officials were initially received with skepticism. "When I spoke to the EIS [Epidemic Intelligence Service] officer, he sounded surprised and just said, 'OK, we'll pick up the culture.'" The EIS was established in 1951, during the Korean War, when the United States worried that the Soviet Union might attack with biological weapons. An arm of the CDC, it was to serve as an early warning system against germ warfare. Through the years its investigators determined that several disease outbreaks were natural and not man-made. Now, after 50 years the EIS was dealing with something different. The country was facing a true bioterrorism siege.

As Quentzel began dealing with the outside authorities, Barakat thought, "Thank God he'll be taking care of the phone. For me I was focused on wanting to help the patient."

On Monday, Ottilie Lundgren was intubated, and that afternoon her chest was drained. "I was optimistic even until Tuesday," Barakat said, "but then she developed multiorgan failure." About the same time, on Tuesday, November 20, Quentzel heard from the EIS that Ottilie's blood samples had been confirmed for anthrax both by the state and CDC laboratories. By then it was clear there would be no recovery. Ottilie Lundgren died the next day.

By the end of the following week, investigators had taken 83 swabs and samples at Ottilie's home and scores more at places she had recently visited, including the Nu Look Hair Salon, the town library, Fritz's Snack Bar, and Immanuel Lutheran Church a mile from her house. Early testing also included 29 samples from the post office in Seymour that serviced her home and 117 samples at the large postal distribution center in Wallingford. All tested negative for anthrax.

Investigators were becoming desperate for clues. The land around Lundgren's house was scoured for anthrax spores. But, as with her mail, furniture, clothes, and other worldly goods, there was no sign of the bacterium. As with Kathy Nguyen, investigators wondered if she had somehow come into contact with a perpetrator posing as a friend. "Did she have any boyfriends?" an FBI agent asked Robin Shaw, Ottilie's hairdresser and friend. "No," said Robin, who was both startled and amused by the question. At age 94? Robin was astonished by how far the authorities were trying to reach.

In later weeks, a connection to the mail seemed more probable if not conclusive. Additional testing showed the presence of anthrax spores on four of the 13 high-speed sorters at the Wallingford center and trace amounts at the post office in Seymour. About a thousand postal employees in the area had been started on preventive antibiotic treatment. A computer printout showed that a letter that went to the Wallingford facility had come from the Hamilton center in New Jersey. The letter had been sorted 15 seconds after the Leahy anthrax letter and on the same machine. Addressed to Jack Farkas at 88 Great Hill Road in Seymour, the letter was found at his house and proved to have a trace of anthrax. No one in the Farkas home, which was 4 miles from Ottilie Lundgren's, had become infected. But the presence of the letter heightened the possibility that cross-contaminated mail might have reached Ottilie's home.

In March 2002, James Hadler, the Connecticut state epidemiologist, revealed that the contamination at the Wallingford center had been greater than previously suggested. Although three of the machines had small numbers of spores, the fourth was "heavily contaminated [with] approximately 3 million spores, roughly translated into 600 infectious doses." Hadler believes the evidence is "strongly suggestive" that Ottilie was exposed through the mail. He says that 29 letters were recovered from her home. Six were first class letters that had been "cleanly opened along the top border" and had been processed only in Connecticut. But of 23 items of bulk mail, "all were torn in half and had been found in her trash." All 29 pieces cultured negative for anthrax. Still, he believed that Ottilie was exposed to cross-contaminated bulk mail. How could this have occurred? Hadler conjectured:

> Well, it's possible that a load of cross-contaminated bulk mail from New Jersey was initially sorted on the heavily contaminated machine in the Connecticut distribution facility, resulting in widespread contamination of that machine. . . . One of these pieces of mail, or possibly another that was cross-contaminated on this machine, then contaminated the sorting machine for her postal route before reaching her home. She was exposed to airborne spores released when she tore this piece of cross-contaminated mail in half—something, again, she only did for bulk mail, not for first class mail.

Fear of anthrax had become compounded by the uncertainty of how Kathy Nguyen and Ottilie Lundgren had become infected. Could the source of their exposures have been so minute as to go undetected despite all the sweeps for evidence? Quite possibly. According to Donald Mayo, who worked with the Connecticut Public Health Laboratories during the anthrax outbreak, "You know, unless you swab every square inch, of course you may miss some spores. You cannot be sure that you've gotten every last one."

The actual number of spores necessary to cause infection remains disputed. Before 2001, animal experiments and sparse information from human experience suggested that a person who inhaled 8,000 to 10,000 spores had a 50 percent chance of becoming lethally infected. After the outbreak, however, the issue seemed more murky. The "preferred interpretation," according to James Hadler, is that Kathy Nguyen and Ottilie Lundgren received exposures from "small numbers of spores."

Jeffrey Koplan, then the CDC director, was skeptical. "It would take a lot more than a few spores to cause inhalation anthrax," he said. But Matthew Meselson, a Harvard microbiologist, disagreed: "There is no justification for assuming there is any threshold at all." Although the chance is small, "a single organism has a chance of initiating infection."

In 2001 the Defense Research Establishment Suffield (DRES), a Canadian government agency, produced a study that was remarkable in its timing. Titled *Risk Assessment of Anthrax Letters*, the study had been undertaken because "no experimental studies on which to base a realistic assessment of the threat posed by . . . 'anthrax letters' could be found." The report of the study was dated September 2001, the very month the first anthrax letters were mailed in the United States. The impetus of the study was a letter received by a government office earlier in the year claiming to contain anthrax. That threat proved to be a hoax. But it inspired DRES to conduct a series of experiments.

Each trial involved placing a letter and a quantity of bacteria, ranging from 0.1 to 1 gram, in an envelope. After sealing the envelope with the flap's adhesive, it was shaken and brought into a 10 × 18 foot chamber. There a tester in mask and protective outerwear "played the role of mail clerk." He tore open the envelope and removed the letter. The relatively harmless bacteria (a simulant of anthrax bugs called *Bacillus globigii*) floated out. Culture plates, which could grow colonies from bacteria that settled on them, had been placed around the room to see how far the germs spread. The results were stunning:

> Before starting the experiments, it was assumed the opening of an envelope constituted a very "passive" form of dissemination that would produce minimum aerosolization of the BG spores. . . . This assumption proved incorrect. . . .The high concentration and rapidity with which the aerosol spread to the other end of the chamber [were] also unexpected. The very heavy contamination on the back and front of clothing worn by the subject [tester] was also unexpected.

Within 45 seconds of the time of release, bacteria had spread everywhere, including onto the breathing filter of the tester's mask. The report suggests that the study's "dramatic results" demanded better preparation for an attack with germs sent by mail. "It is only a matter of time until a real 'anthrax letter' arrives in some mail room," the report concluded.

As eminently correct as that observation proved to be, another of its admonitions fell short. Failure to seal an envelope's corners, according to the report, "could also pose a threat to individuals in the mail handling system." True. But the implication is that bacteria would not escape from a well-sealed envelope. A month later, during the anthrax letter crisis, investigators realized that the 1- to 3-micron spore particles were leaking not through leaky flaps but through 20-micron pores in the paper envelope.

In its concluding paragraph the DRES report noted that similar experiments had recently been conducted by another agency in Canada, the Ottawa-Carleton First Responder Group. In that study, titled *Investigation on the Dispersal Patterns of Contaminants in Letters,* the Ottawa-Carleton group placed fingerprint powder in envelopes. The results were similar. Even when the envelopes were not physically opened, powder leaked out and caused "contamination in the immediate area."

The risk of anthrax in the mail was also assessed in William Patrick's 1999 study for the Scientific Applications International Corporation. The report is not publicly available, but Patrick shared information about it with William Broad, a *New York Times* reporter. Patrick indicated that a puff of aerosol bacteria emerging from an envelope could be lethal. But there apparently was no indication in his report, or in the two Canadian studies, that cross-contamination could magnify the potential for death.

Just how long would a contaminated mailbox retain spores? Kathy Nguyen's and Ottilie Lundgren's mailboxes tested negative for anthrax. But the tests were undertaken days after the women's diagnoses and probably weeks after an infected mail item would have arrived in their boxes. If a slightly contaminated letter left spores in a mailbox, how long might it take for subsequent deliveries and retrievals to clear the spores from the box?

Late in 2002, I conducted a study to help answer that question. Published in the *Journal of Public Health Management and Practice* (September 2003), it is titled "Persistence of a Mock Bio-agent in Cross-Contaminated Mail and Mailboxes." The study included three trials, each conducted in the same manner. A trial would begin with the placement of 1 gram of a mock biological agent and a sheet of paper in a prestamped postal envelope, much like those used in the real anthrax mailings. The agent, called "Glo Germ," is a fine white powder composed of 5-micron particles that approximate the size of anthrax spores. The particles are fluorescent and strikingly visible under an ultraviolet light. Glo Germ is used to check hygienic and infection control practices in hospitals and among food handlers. It has also reportedly been used by the U.S. Army as a tracer in mock germ warfare tests.

After sealing the envelope flap and its corners with tape (as was done with the real anthrax letters), the envelope is placed in two Ziploc plastic bags, one inside the other. The bags are shaken and squeezed, to simulate the turbulence caused by a mail sorting machine. The letter is then removed and replaced with six uncontaminated letters. Again, the bags are shaken. The six letters are removed from the bags and placed in a rural mailbox much like the one used by Ottilie Lundgren. After 3 minutes, the letters are removed from the mailbox. Fluorescent particles on the letters and on the floor of the mailbox are counted.

The exercise is repeated: Nine more deposits and retrievals in the same mailbox are performed, each with a new batch of six uncontaminated letters. After each "mailing," the letters and the floor of the mailbox are examined for the presence of particles.

The first six letters in each trial are the only ones that have had direct contact with an originally contaminated letter. Not surprisingly, the total number of particles in each of these initial batches was high—exceeding 50 particles. In all three trials, the number of particles—that is, the level of contamination—decreased in the course of the 10 mailings.

Particle counts on the floor of a mailbox after the first mailing varied in each trial. In the first trial the figure exceeded 50, in the second it was 32, and in the third it was 24. The enumeration of particles was in part subjective because some of the fluorescent spots appeared as smudges rather than discreet points. (A smudge may have represented several particles or otherwise been an artifact, but each was recorded as one particle.) Still, the trend was clear. After the fourth or fifth mailing, traces of fluorescent particles were largely absent from the letters. While traces in the mailboxes themselves persisted somewhat longer, their numbers also declined with successive mailings. Indeed, in the first trial, after the tenth mailing, no particles were visible in the mailbox. In the other two trials, after 10 mailings, two particles remained in each.

This exercise demonstrates that successive placements and removals of letters in a mailbox may cause a decline in the number of residual bioagent particles. It supports the mail-related explanation for the anthrax infections of Kathy Nguyen and Ottilie Lundgren. The women may well have been exposed to anthrax in cross-contaminated mail despite the apparent absence of spores in their mailboxes or among their personal belongings weeks afterward. The pool of spores that they were exposed to may have been very small. And the few spores that remained after exposure could have been cleared away during subsequent contact with uncontaminated items.

These observations provide a sense of both comfort and unease. It is clear that cross-contamination of mail sharply declines with successive placements and retrievals in a mailbox. After just a few mailings, even though some bio-agent might still be found in a mailbox, newly placed letters remain largely free of particles. Moreover, the box itself becomes decreasingly contaminated during successive mailings. Exactly where all the particles end up is uncertain. But most do not attach to the letters and are probably swept into the air as the letters are withdrawn. Assuming small numbers, they could settle into the ground beyond detection or, if indoors, be swept from surfaces during later cleaning.

Still, the fact that anthrax spores are very durable, and the chance that some people are susceptible to infection from a very small amount, is disturbing. Tiny quantities of spores from the anthrax letters could have found niches in homes around the country. While most people would appear not to be vulnerable, the fate of Kathy Nguyen and Ottilie Lundgren reminds us that some people might well be. Years from now a few anthrax spores on a piece of long-saved mail, or on another cross-contaminated surface, might yet infect a vulnerable person. The legacy of the 2001 anthrax letters could linger long into the future.

The anthrax letters generated a host of unexpected findings and changed assumptions. Conventional wisdom about anthrax infection before the attacks was subjected to reassessment in many areas—diagnostics, surveillance, gravity of the disease, manner of treatment, effectiveness of transmission by mail. Perhaps the greatest overall effect has been the nation's altered mind-set about bioterrorism. In consequence of the actual experience, almost no one doubts the likelihood that there will be more such attacks. Complacency has given way to concern. And no one on the planet better exemplifies the changed manner of thinking than Dr. Donald Henderson.

CHAPTER
SIX

D.A. HENDERSON, THE CDC, AND THE NEW MIND-SET

A few blocks from Charles Street in Baltimore, near the undergraduate campus of Johns Hopkins University, the neighbor hood abounds with handsome single-family brick homes. On Sunday afternoons in September, residents are usually relaxing, perhaps watching football on TV. But on Sunday, September 16, 2001, Dr. Donald Ainslee Henderson, like many of his neighbors, was still thinking about the terrorist attack 5 days earlier or, more specifically in his case, about what might happen next. "Tara, Tom, and I had been talking since right after September 11," he says, "and we felt there was likely to be another event."

D.A., as Henderson likes to be called, was referring to Dr. Tara O'Toole and Dr. Thomas Inglesby, director and deputy director, respectively of the Center for Civilian Biodefense Strategies. Not surprisingly, the likely "event" they had in mind was one involving a germ weapon. The center is a think tank at Johns Hopkins that Henderson founded in 1998 with Dr. John Bartlett, chief of infectious diseases at the Hopkins Medical School. Its avowed aim is to help develop policies "to protect the civilian population from bioterrorism." Among the growing number of biodefense centers at academic institutions—New Jersey's University of Medicine and Dentistry, George Mason University, St. Louis University, and the Universities of Louisville, Minnesota, South Florida, and Texas—the one at Hopkins stands apart. Not only was it the first, it also was the beneficiary of Henderson's unusual credentials.

For 14 years, beginning in 1977, Henderson served as dean of the Johns Hopkins School of Public Health. But before then he had already become a legendary figure in public health. From 1966 to 1977 he directed the World Health Organization's campaign to eradicate smallpox. His last year as director was the last year that a naturally occurring case of smallpox occurred anywhere in the world.

As he sat at home that Sunday afternoon, the phone rang. "Dr. Henderson? This is Eric Noji at the Department of Health and Human Services [HHS]. Secretary Tommy Thompson asked me to call." Dr. Noji, a specialist in "disaster medicine," had recently come up from the Centers for Disease Control and Prevention to help develop an Emergency Command Center in the secretary's office. "Secretary Thompson would like you to come over for a meeting at 7 o'clock." Henderson had worked in the previous administration as deputy assistant secretary at HHS under Donna Shelala. He had never met Thompson and much appreciated the invitation. "Seven tomorrow morning? Or do you mean tomorrow evening?" Henderson asked.

"No, no. Tonight," Noji answered.

Henderson caught the sense of urgency. "Yes, of course I'll be there."

Soon after, he was driving south on Interstate 95 for the hour-long trip to Washington, D.C. Once in the city, he worked his way quickly along Independence Avenue—Sunday traffic is light. As he approached 3rd Street, the open mall on the left offered a stunning view of the Capitol building. To the right, still on Independence Avenue, stood the Hubert H. Humphrey Building, the headquarters of HHS. The seven-story structure, constructed 25 years earlier, is covered with recessed windows that give the appearance of a giant waffle.

Henderson parked on the street and walked to the building. As he entered the lobby, high on the wall to his left he could see a portrait of Humphrey and a gold-leaf inscription. The text declares that the manner in which a government treats children, the elderly, the sick, and the needy is, in Humphrey's words, "the model test of a government." Six floors up, Henderson got off the elevator and turned right. Before him was a glass partition beyond which lay the red-carpeted suite of Secretary Thompson.

He was escorted into Deputy Secretary Claude Allen's conference room. Propped up on a rectangular conference table were a computer and a monitor—a rudimentary command center soon to be vastly expanded. A couch stood in front of one wall, and soft chairs were off to the side. "I'm very glad you're here," Secretary Thompson said to Henderson. They were joined by Allen, Noji, and Scott Lillibridge, who had just been appointed director of the command center.

Thompson began the discussion by inviting comments about the current situation. The general feeling was that there would be a terrorism sequel to the September 11 attacks. But what form would it take? Henderson recalled the moment:

> We sort of worked our way through the discussion. Doing something with an airplane again was going to be much harder now than it had been, we decided. I think we all came to the conclusion that it could very well be a biological event. And it was quite apparent to me that this was Thompson's view too.

The secretary rose from his seat and paced back and forth. "He was obviously extremely distressed," Henderson said. Then, referring to his contacts in the White House, Thompson said, "They just don't understand."

"What don't they understand?" Henderson asked.

"Biological weapons," Thompson answered. Secretary Thompson repeated his concern, emphasizing that the country was "unprepared, grossly unprepared for a biological attack."

During the next weeks, Henderson was repeatedly invited back to confer about the possibility of further terrorism. Then, on October 4, Bob Stevens, in Florida, was diagnosed with anthrax. The same day federal and state authorities publicly dismissed bioterrorism as a likely cause of Stevens's illness. Thompson himself emphasized that the case was an isolated incident. He implied that Stevens might have contracted the disease from water: "We do know that he drank water out of a stream when he was traveling to North Carolina last week." When Henderson heard this, he was mystified.

Thompson's statement was uninformed. Water is not known to convey anthrax, and drinking would be unlikely to cause the inhalation form of a disease. Having minimized the possibility of bioterrorism, the secretary was later criticized as being naïve, as not appreciating the serious implications of the incident. In fact, according to Henderson, Thompson's concerns were very real. Henderson realizes this view seems "contrary to what came out when Thompson got on television and assured everybody that everything was in great shape." But Henderson is convinced that

Thompson did not believe everything he was saying publicly: "He did what I have seen happen in many disease outbreaks. The political figure gets out in front and says, 'Everything is under control. Please relax.' You know, to calm everybody down. And that isn't necessarily the right thing to do."

What would Henderson have advised the secretary to say that first day?

> To acknowledge that there is a problem. That there's a lot of work to be done. To say that he will keep in close touch with the public. In other words, be open about it. But the tendency is to say, "Everything is in good order. We've got everything under control, so don't worry." I think that's wrong.

A few days later the presence of anthrax spores was confirmed on Stevens's computer keyboard and elsewhere in the American Media building, where he worked. By then, as anxiety heightened about the source of the anthrax, Henderson had again been summoned by Thompson. From that time forward, Henderson began spending most of every day on the sixth floor of the Humphrey Building. On November 1 the Secretary announced the establishment of the Office of Public Health Preparedness, which would be organized and directed by Dr. Henderson.

Henderson, then 73, had agreed to serve for 6 months. In May 2002 he stepped down and was succeeded by Jerome Hauer, who previously had overseen the Office of Emergency Management for New York City. Henderson continued on as Secretary Thompson's principal science adviser. His new schedule still had him catching the 7:10 a.m. train out of Baltimore. But now he was usually home by early evening, only an 11-hour work-travel day. Not like his half year as director when he was working "day in, day out, evenings, without letup," he said.

D.A. Henderson grew up in Lakewood, Ohio, outside Cleveland. He attended Oberlin College and in 1954 graduated from medical school at the University of Rochester. He had planned to become a cardiologist but, facing the military draft, decided to volunteer for 2 years of government service. He applied to the Communicable Disease Center (forerunner to today's CDC) and between 1955 and 1957 was assigned to its Epidemic Intelligence Service.

The EIS had been created out of concern that American troops might be exposed to germ weapons.

By the time Henderson joined the EIS, the Korean War was over and "interest in biological weapons had disappeared," he said. (In fact, between 1949 and 1969 the United States was developing and stockpiling biological weapons, as were the Soviets, but the public knew little about those programs.) Henderson's interest shifted from cardiology to epidemiology and infectious diseases, and he went on to receive a master's degree in public health from Johns Hopkins. Then in 1961 he returned to the CDC as director of disease surveillance. More than three decades would pass before he and the CDC would again be thinking about bioweapons and bioterrorism. But a year after the anthrax attack, he was thinking about little else.

At 11 a.m. on a clear day in November 2002, following a presentation on smallpox vaccine at the Chemical and Biological Arms Control Institute in Washington, D.C., Henderson arrived at the Humphrey Building. As he got off the elevator on the sixth floor, he turned left. Down the hall and around the corner, he reached a door marked 636G. He swiped a card to gain entry to the Office of Public Health Preparedness. Jerry Hauer, then the assistant secretary for public health preparedness, waved a greeting from his spacious office on the left. At the end of a short corridor, Henderson entered his own office and hung his dark suit jacket on a coat rack. The bookshelves were nearly empty and the absence of pictures and decorations suggested a condition of impermanence. A large American flag hung from a pole next to his desk. He glanced at a white table fan. "Circulation is not that good here on warm days," he chuckled.

Even the hottest days are more comfortable than those he had spent in the tropics in the 1960s and 1970s. As director of the Global Smallpox Eradication Campaign, he traveled from his comfortable Geneva headquarters to spare regional offices around the world—Senegal, the Ivory Coast, Indonesia, India. Etched in his memory are the oppressive heat and humidity. "When I would try to write my reports, sweat would pour down my arm." He raised his right arm and went through the motions of a long-ago ordeal. "I'd wrap my arm with a towel and prop my arm up to stop the flow." The reports were often drenched anyway.

Henderson is a big man. His 6-foot-2-inch frame remains imposing even as he sinks into a seat at the conference table near his desk. When he leans back and clasps his hands behind his head, his tie and suspenders shift forward. A smile accentuates the crow's feet behind his gold-rimmed glasses. I asked how he came to head the smallpox eradication effort. "In 1961, I had returned to the CDC and become head of the surveillance program. Three years later we began assisting in a measles vaccination campaign in six countries in Western and Central Africa. Then we decided to include smallpox vaccinations as well." During the next few years, the campaign against measles made little headway, but the smallpox effort was more successful. As the number of new cases began to fall in Guinea, Nigeria, Sierra Leone, and elsewhere, the program was expanded to include 20 countries. Meanwhile, the 20th anniversary of the United Nations was approaching and the organization declared in 1965 that the following year would be the "International Cooperation Year."

"And then there came a surprise," Henderson said. His deep baritone rumbled like a low-decibel motor: "President Johnson decided he wanted the U.S. to take an initiative showing international cooperation. Some people from the Public Health Service suggested an effort to eradicate smallpox in the 20 Central and West African countries. And the president decided to go with that."

The initiative soon expanded to the breathtaking goal of eliminating smallpox everywhere. The idea sounded especially appealing because of the unusual devastation wrought by the disease. During the first 10 days of infection a person shows no sign of illness. But by the end of the second week, fever is accompanied by crushing headache and backache. Rashes begin to appear on the face, trunk, and limbs. They erupt into boils that turn the body into a terrain of pus-filled mounds. As many as 30 percent of smallpox victims die.

The fatality rate for smallpox is lower than for inhalation anthrax. But unlike anthrax, smallpox can be transmitted from person to person. If a boil is scratched open, the oozing pus that glazes the surface can infect anyone who touches it. And with every cough a victim releases viruses into the air that can infect someone in the vicinity. Survivors of smallpox are forever stamped with deep scars, the remnants of the ubiquitous postules.

The idea to rid the world of smallpox was not entirely new. Resolutions to eliminate smallpox as well as malaria and yellow fever had been endorsed by the WHO. But actual efforts, such as they were, had failed. The extraordinary ambition of such an undertaking competed with the cold reality that no disease had ever been rendered extinct through human effort. Yet certain characteristics of smallpox, which is caused by the *variola* virus, seemed to make it amenable.

In the first place, the only natural host in which the *variola* virus can survive is a human being. Some killer microorganisms, like anthrax bacteria, can thrive in animals or lay dormant in the soil for years. But in the case of smallpox, if human infection is eliminated, the virus disappears from any natural setting.

In addition, a single vaccination against smallpox provides immunity for years. So vaccinating enough people should mean that the virus at some point would have no natural habitat in which to "live."

Still, not everyone was happy about the proposal for a global initiative. "D.A., I'm furious," complained Alexander Langmuir, who as chief of epidemiology at the CDC was Henderson's boss. Langmuir believed the plan was so ambitious that it would drain resources from his own CDC programs. "It will destroy the Epidemiological Intelligence Service," he railed.

Other prominent figures were convinced that an eradication program could not succeed in any case. One was Marcolino Candau, the director-general of the WHO. He believed that success would require vaccinating every last person in the world, an impossible task. But since the campaign would be global, the proposal was put in the lap of the World Health Assembly, the deliberative body of the WHO, which meets annually. In May 1966, after several days of debate, the 120-odd member states voted to support an eradication campaign by a margin of two votes.

Candau faulted the United States for what he was sure would be a failed program. "He wanted an American running the program, so that when it went down the tubes the U.S. could be blamed," said Henderson. Since Henderson had been leading the CDC's measles-smallpox program in Africa, Candau asked that he be appointed director of the new global initiative. Henderson protested:

> I had too much that I was attending to already. Also, I knew that whatever could be achieved would have to depend on consensus and persuasion. I had real doubts that you could eradicate a disease given the political realities of the organization.

Did Henderson think it medically possible to eradicate smallpox?

> From a technical standpoint, yes. Bear in mind, we knew the U.S. was free of smallpox. Europe was free of the disease, as were a number of countries in North Africa. The demonstration was there. But administratively, I had serious doubts. But I was told that either I went or I would have to resign. I liked the Public Health Service. So we struck a deal. I would get the program started and come back after 18 months.

When the program began in 1967, smallpox was endemic in more than 30 countries. The WHO eradication campaign was to take 10 years. By mid-1968, after the first 18 months had passed, Henderson felt he had made progress. He decided to remain at the helm a bit longer. In the end, he did not step down until 1977, the year that the last case of endemic smallpox was found.

At the outset, Henderson had no masterplan. But as the program went from region to region, cumulative efforts led to successes. "Our last case in West Africa was in Nigeria in 1970, then Brazil in 1971, Indonesia in 1972." By then the program was being phased in elsewhere—in Afghanistan, Bangladesh, India. Beyond earlier efforts to simply immunize as many people in a region as possible, the strategy came to be based on disease surveillance. Local health authorities would deliberately seek to learn about any outbreak of smallpox. When a case was detected, they would isolate the infected patient and vaccinate everyone the patient had been in contact with. This manner of containment became known as the "ring" approach, by which a circle of immunized people would become a barrier to new infections beyond their perimeter.

The strategy ultimately proved successful. But the effort could be harrowing. In *Smallpox and Its Eradication*, a book produced by the WHO (1988), Henderson listed the formidable obstacles faced by workers in the field: famine, flood, epidemic cholera and other diseases, and frequent changes of governments whose interest in cooperating varied. The most dangerous challenge came from civil wars and local conflicts.

Dr. Donald Millar, who led the CDC contribution to the global program, was engaged in field operations in several countries. "I was placed under house arrest during coups seven times," he recalled in a genteel Virginia drawl. "Togo, Ghana, Benin. In Nigeria the civil war was fought while we were there." Although he was never injured, he often worried.

> You never knew when you approached one of those roadblocks if it was going to be a major hassle. Is somebody going to get killed here, or what? Some of our people were roughed up at these checkpoints. I

> was with Dr. Margaret Grisby, who had been a professor of medicine at Howard University, when she got hit with a rifle butt. There was no provocation. Maybe they didn't like her looks. Who knows?

As Henderson observed, the only power possessed by WHO was that of "moral suasion." From the beginning, Henderson realized that the country that posed the greatest challenge to the eradication effort would be India. "Of the 1.1 billion people living in areas which had endemic smallpox in 1967, 513 million lived in India," he noted. In fact, the *variola* virus is thought to have evolved into a human killer on the Indian subcontinent. In the mid-20th century, half of the world's 2 million annual smallpox fatalities were in India. The country's large land area, its poor system of health reporting, and its huge mobile population—-millions on religious pilgrimages and other travels—had undermined previous efforts to control the disease there.

But with the participation of local and regional health authorities, in 1973 the WHO intensified its outreach efforts. Workers fanned out to villages and markets to find recent cases. Photographs of a smallpox patient and promises of reward for finding other patients were posted in markets and teashops. Workers stationed at market entries would ask people if they knew of any cases and which village they were from. In one district, in Assam, 34 of 695 villages were found to have had cases of smallpox. When a case was identified, quick action followed, as described in *Smallpox and Its Eradication*:

> "Watchguards" were hired to stay at each infected house to prevent the patient from leaving and to vaccinate anyone who could not be dissuaded from visiting. . . . When it was found, in some areas, that visitors avoided the watchguard by entering through a back door, the back entrance was barricaded.

At the same time, "vaccinators" would try to reach every person within a 1-mile radius of an outbreak. Just how difficult the effort could be was exemplified in Puri village, a pilgrimage site of the Jain religion. In December 1974, at the height of the pilgrimage season, 40 Puri households had become infected. Henderson noted that many Jains resisted vaccinations even after their religious leader "reluctantly agreed" to recommend them. In consequence, the entire village was quarantined by military police, and Jains were kept from the area unless they were vaccinated. By that time a pattern had become established. As described in the book:

> Whenever an outbreak was discovered, 20-25 vaccinators were dispatched to the infected village; containment vaccination was completed within 48 hours; 24-hour watchguards were posted at every infected household; and food was brought in to ensure household quarantine.

By the beginning of 1975, smallpox remained in only 285 of India's 575,000 villages. But challenges to the eradication efforts had not eased. Here are excerpts from a report at that time by a region supervisor in Bihar, an Indian state on the northern border near Nepal:

> We have the misfortune to have to inform you of a new case of smallpox in the Painathi outbreak, a 4-month-old unvaccinated male. . . . The family had been resistant and uncooperative from the start. . . . After a rumor reached Dr. Khan, who had been staying in the village, he had to use a trick to gain entrance into the house. He asked for a glass of water and this was denied. He knew by custom that they had a case of smallpox inside the house because nothing can be given when a case of smallpox is in the house of a member of this religious sect.

The most stunning sentence of the report was: "Vaccination was possible only when we climbed over the compound walls and forcibly inoculated each family member." Such extraordinary efforts led to continued progress. The last case of smallpox in India was identified in May 1975. She was Saiban Bibi, 30, a homeless beggar who had lived on a railway platform in Assam. Bibi was brought to a local hospital where four guards were assigned to make sure that she remained isolated. During the next 2 weeks, searches and surveillance were intensified at locations that were serviced by the railroad. No additional cases were found.

India celebrates Independence Day on August 15. In 1975 that day included special recognition of India's freedom from smallpox. "It was a great event," Henderson recalled, relishing the memory. "There were parades, the WHO staff was there, the Indian health workers. We were given plaques and medals, and we were draped with garlands."

More than 100,000 workers, including interviewers, vaccinators, and watchguards, had participated in the country's eradication program. The monumental success was captured by a simple chart that appeared 2 years later. In 1974 the number of smallpox cases in India was listed as 188,003; in 1975 the figure was 1,436; in 1976 it was zero.

The eradication campaign continued in other countries, but after India, Henderson said, “we felt we were going to make it.” Two years later that expectation was fulfilled. On October 26, 1977, in Somalia, a thin 23-year-old cook was diagnosed with smallpox, the last naturally occurring case on earth. “And so the final chapter was written,” Henderson later recorded. “Ali Maow Maalin represented the last case of smallpox in a continuing chain of transmission extending back at least 3000 years.”

A sad postscript occurred in 1978 when Janet Parker, a medical photographer at the University of Birmingham Medical School, died of smallpox. She had become infected with virus that had escaped from a laboratory one floor below the one on which she worked. No cases appeared after that, and on May 8, 1980, the World Health Assembly declared that the battle against smallpox was over and vaccinations would no longer be necessary.

Shortly before the 1991 Persian Gulf War, Henderson ended his tenure as dean of the Johns Hopkins School of Public Health. He had accepted an appointment as associate director of the Office of Science and Technology Policy, an advisory body to the president of the United States. The possibility that Iraq might use biological weapons was being discussed, “but we weren’t overly concerned,” he said. The United States did not yet understand the extent of Iraq’s germ warfare program. Henderson continued in government service under the Clinton administration, becoming deputy assistant secretary for health at HHS. But he left that post in 1995, a year he describes as a watershed.

In 1995, Saddam Hussein’s son-in-law, Hussein Kamel Hassan, defected to Jordan. Kamel had overseen Iraq’s development of weapons of mass destruction. His defection prompted the Iraqis to admit to a much larger biological weapons program than they had been acknowledging to United Nations weapons inspectors. Around the same time, the public was hearing more from Ken Alibek and other Soviet defectors about the expansive biological program they had run in the former Soviet Union. It was also the year that Aum Shinrikyo, a Japanese cult, released sarin nerve agent in the Tokyo subway. Although sarin is a chemical, the incident suggested one way in which a biological agent might easily be unleashed. “This was when I began to become concerned. But even then, to tell you the truth, I didn't really do much about it," Henderson acknowledged.

His worries ripened in 1997 after passage of the Nunn-Lugar-Domenici Domestic Preparedness Act (named for Senators Sam Nunn, Richard Lugar, and Peter Domenici). The legislation provided $36 million to the Department of Defense and $6.6 million to the Public Health Service to train police, fire, and other responders in 120 cities to deal with mass casualty attacks. Henderson, however, was not pleased: "It became apparent to me that the responses were being crafted by police and chemical and military people. It was all focused on 'bang' or on gas release. There was very little attention being paid to biological."

By then he had already had several conversations about the threat of bioterrorism with Michael Osterholm, a public health colleague. Osterholm, the state epidemiologist for Minnesota, had been speaking publicly on the subject for a few years. "Mike was ahead of me in perceiving that there was a real problem," Henderson said. Osterholm recalled a 1993 meeting at which Soviet defectors said their former regime had produced tons of anthrax and smallpox to be used as bioweapons. "That briefing created a visceral memory for me," said Osterholm. He became increasingly convinced that the United States was unprepared for a biological attack, but he felt frustrated because he could arouse little interest in others.

Henderson came to share Osterholm's worries, especially about the possible reemergence of smallpox. Still, he felt that public discussion would not be helpful. That all changed toward the end of 1997, when he finally agreed to appear on a panel to discuss bioterrorism. "It was September 1997," Henderson recalled, "at the opening session of the annual meeting of the Infectious Diseases Society of America—IDSA." In addition to Osterholm and Henderson, another panelist was Richard Preston, whose fictional thriller on bioterrorism, *The Cobra Event*, was about to be published. As they sat on the stage, the panelists were amazed. Henderson turned to a wide-eyed Osterholm on his left and said, "Can you believe this? The hall is packed, standing room only." "There's gotta be more than 2,000 people here," Osterholm replied. John Bartlett, who was president of IDSA, later told Henderson that throughout the remaining days of the conference the "discussion came back again and again to the risk of bioterrorism." Henderson had begun his remarks with an acknowledgment of his conversion:

> Until recently, I had felt it unwise to publicize the subject because of concern that it might entice someone to undertake dangerous, perhaps catastrophic experiments. However, events of the past 12 to 18 months have made it clear that likely perpetrators already envisage every agenda one could possibly imagine.

After his appearance on that panel, Henderson was flooded with speaking invitations from groups of all types, he says, including hospital staffs, policy associations, and the governing board of the National Academy of Sciences. His message was that too little was being done to prepare for a bioterrorism attack. The following year he invited 24 people to his fledgling biodefense center at Johns Hopkins to develop policy initiatives. "Some of the people were in government, some were academics, some public health doctors. We sat down in the fall of 1998 and asked 'What are the organisms we're most concerned about?'"

Until then, more than 30 potential microorganisms and toxins had been listed by the U.S. Army and NATO as potential biowarfare agents. The Hopkins working group determined that six of them were the most likely to be used and deserved particular attention. Designated "Category A" agents, at the top of the list were the bugs that cause smallpox and anthrax and then plague, botulinum toxin, tularemia, and viral hemorrhagic fevers (such as Ebola or Marburg).

The heightened concern about smallpox was only recent. The last case of smallpox in the United States had been in 1949, and in 1972 routine vaccinations in this country were ended. (By then the disease was absent in the Western Hemisphere. The occasional serious side effects from a vaccination no longer seemed worth risking.) In 1984, four years after its declaration of global eradication, the WHO authorized only two laboratories in the world to retain stocks of *variola* virus. One was at the CDC in Atlanta and the other at the Research Institute for Viral Preparations in Moscow.

Thus, on the assumption that the virus was safely locked away, U.S. authorities considered smallpox an unlikely bioweapon. As recently as 1994, a publication by the U.S. Army Medical Research Institute of Infectious Diseases had failed to mention smallpox among 12 biological threat agents. But the discovery that the Soviets had been illegally producing the virus, and suspicions that other countries (notably Iraq and North Korea) might be doing so, prompted a reassessment. The push from Henderson, Osterholm, and others resulted in a reordered sense of risk. Almost all Americans were now deemed vulnerable to smallpox since none had received vaccinations in the past 30 years.

Meanwhile, Henderson had been unsuccessfully seeking funds for his new biodefense center at Johns Hopkins. The center's start-up money, a $1 million congressional appropriation sponsored by Senator Barbara Mikulski, was running out. He spoke ruefully about his drawerful of rejections:

> We went to the Gates Foundation, we went to Robert Wood Johnson, we went everywhere. Their attitude was, "I think our trustees would find it a little awkward to support something on biological weapons, you know, difficult for our image." Then we went to the ones that normally support arms control. And their response was, "We never support work in the area of public health and medicine." Nobody wanted to support us for a center. Nobody.

Before giving up, Henderson tried one more possibility—Ralph Gomory, whom he knew from meetings of the National Academy of Sciences. (Gomory is a member of the Academy and Henderson a member of the allied Institute of Medicine.) Gomory, a mathematician, is president of the Alfred P. Sloan Foundation. As Henderson later learned, he had an abiding concern about bioterrorism.

Gomory is a slightly built man with a businesslike demeanor. Early in 2000 he had ordered disaster kits for everyone who worked at Sloan. Every employee now had a kit at his desk that included a whistle, flashlight, gas mask with filter tips, and more. In 2002, during a conference on bioterrorism sponsored by the foundation, Gomory stood to the side of the room and said he wanted to show me something. He reached into his left pants pocket and pulled out a blue cloth protective mask. "I carry this all the time," he said, his brows furrowed with a sense of gravity. "It will be helpful against either a biological or chemical agent." Gomory's preoccupation had been Henderson's good fortune.

"D.A., How many foundations have supported you until now?" Gomory asked when Henderson first approached him.

"Precisely none, Ralph."

Henderson was bowled over by Gomery's reaction. "Great," he said, "We're definitely interested. Send us a proposal."

"For how much money?" Henderson asked.

"Why don't you ask for what you need, and we'll cut it back if we think it's too much."

Henderson huddled with his staff at the center and came up with a request for $3.5 million. They sent the proposal in and held their breath. Three weeks later they received an answer. Henderson grinned, "We got a check back for $3.5 million."

Subsequently, Henderson heard from Gomory that the foundation was "absolutely delighted" with the decision to grant the money. It had become clear in Congress and elsewhere, Henderson said, that "if not for the center and for Sloan, we would have been much less well prepared for bioterrorism than we are." During the anthrax incidents, the Hopkins center was a magnet for callers, especially from the media, who were hungry for information. Five of the center's experts were on the phone 10 hours a day, fielding inquiries from the press. Actually, though, the public health backbone of the nation's preparedness for a bioattack is the CDC. The anthrax incidents tested that responsibility as never before.

Stephen A. Morse became associate director of the CDC's bioterrorism preparedness and response program when it was established in 1999. A microbiologist, he had previously been working on the AIDS virus at the CDC. In March 2002 we met over beers at the Regency Hyatt Hotel in Atlanta. It was between sessions of a conference on emerging infectious diseases, sponsored by the CDC and the American Society for Microbiology. Richard Kellogg, who coordinates the CDC's bioterrorism laboratory response network, joined us.

I asked what life was like at CDC the previous fall. Morse spoke evenly through a close-cropped salt-and-pepper beard. He mentioned a meeting of CDC officials that convened just after Bob Stevens's blood samples were confirmed for anthrax.

> We sort of bandied about the idea that this was a terrorist act. But because there was no direct evidence—no threat letter or anything else, I guess the tone was "Let's wait and see. Let's do an epidemiological investigation and find out if this is just a freak isolated case." I mean, we were already geared up after September 11. There had been concern that there could have been a biological agent on the planes that hit the World Trade Center. We had people on the ground in New York doing surveillance of the hospitals for possible bioterrorism.

Kellogg reflected on their exhausting schedule: "After 9/11, if I'd go home, Stephen would borrow my sleeping bag because he'd be on the late watch. We were all waiting for the next thing to happen." And then the anthrax letters. I asked Kellogg whether he felt traumatized by all this. He shook his head no and in a southern accent that harks back to his Florida childhood answered:

> You're tired is what you are. You get into an initial period where the adrenalin is flowing. But people are human beings, and there is going to be a fatigue factor. I think that was one of the big issues, working 24 hours, 7 days a week. The quality of decision making suffers once people become fatigued. That's something that we're going to have to work through for the future.

Morse and Kellogg were occupied with coordination of laboratory services around the country. By the time of the anthrax letters, the CDC had established a laboratory network in which 80 state and local public health laboratories were participating. These laboratories had received instructions and materials for testing several likely bioterrorism agents. The laboratory in Jacksonville, where Phil Lee confirmed Robert Stevens's anthrax, was part of the network. But all the laboratories, no matter how well instructed, were experiencing a crushing surge in demand. Physicians, hospitals, and law enforcement agencies were sending in so many samples that even the best-prepared laboratories could not keep up.

In October and November 2001, more than 120,000 specimens were tested for anthrax, an average of 2,000 per day; but all told the number of laboratories involved—in the military, the Federal Bureau of Investigation, at CDC, and in the national public health response network—totaled less than 150. Round-the-clock efforts by laboratory personnel were heroic but pointed to the need for a more expansive testing capability. (A year later the number of labs in the national response network alone had increased from 80 to 120.)

I asked Morse and Kellogg if they or their families treated their mail differently during the anthrax scare. "No," Morse answered. "The only thing I did was make sure we had a supply of Cipro in the house." Kellogg did the same. "We already had our own ministockpile in our travel kits. I just gave my wife the packet and said, 'here you are, just in case.'"

Were they ever worried personally? "I'll tell you when I was worried," Morse answered, then shifted the focus back 3 years.

> I got involved in bioterrorism in January 1999 when the program got started and Scott Lillibridge became director. I was the deputy. For the first 6 months I was receiving classified briefings on what the threat was. I was having nightmares those 6 months.

Morse knew that laboratories nationwide lacked assays to test for potential germ weapons—that is, the chemical reagents that could confirm the presence of organisms like anthrax and plague bacteria. "I was so worried that something was going to happen before we got the laboratory response network up and running."

Kellogg joined in:

> See, everybody back then was taking the matter less seriously. We were getting the briefings and we realized how serious it was. So we were working round the clock getting those labs set, getting the funding out there, getting them up to speed, conferences, explaining—all these things.

He offered a half smile: "It was fortunate that the perpetrators waited until we were ready."

An hour into the conversation, Scott Lillibridge joined us at the table. His round face is boyish, a nice match for his dry wit. But his current assignment is deadly serious. For the past year he has been at HHS, as an adviser on bioterrorism to Secretary Thompson. "Actually, now I'm working with D.A. Henderson to move grant money for bioterrorism preparedness programs to the states," he said. I asked Lillibridge what he thought when he learned of the first anthrax case: "My reaction was—bioterrorism. I was thinking, 'Will they be putting it into the air ducts of a federal building?' I was surprised they only used five or six letters."

So what are you worried about now?

> I worry that a threshold has been crossed, and others will be more inclined to do bioterrorism. And I worry about the explosion of new biotechnology and bioscience. I can barely keep up with the new findings. You know, if you just change the DNA code a little bit in one of these organisms, we could be giving the wrong prophylactic medication. Imagine that.

Morse mentioned how surprised everyone was about the method of attack. "We had been thinking about the military mode of transmission of these agents, by an aerosol release." Kellogg and Lillibridge nod in agreement. Like many others, they were shocked to learn how effective the mail could be to deliver the anthrax. Morse shakes his head to lend emphasis as he adds: "We were just fortunate that it was small and that we could learn a lot about inhalation anthrax and how the organisms could be spread."

They all agreed that after the first anthrax case communication between agencies was poor. Some of the leadership at the CDC and the FBI were reluctant to share information with each other. Did the leaders want the limelight for themselves? "Well, for a while, it seemed they all wanted to be president," Kellogg quipped.

Newspapers had reported complaints that the CDC's hotline was providing confusing information. Dr. Daniel Ein, a physician in Washington, D.C., despaired that his inquiries were met with "fumbling" answers. Even communication among CDC investigators was erratic. Epidemiologists in the field despaired that they were getting more information from news reports than from colleagues in Atlanta. The agency also faced a barrage of criticism after the deaths of the two postal workers for not having foreseen that spores could leak from envelopes.

Through it all, CDC Director Jeffrey Koplan insisted that the centers were acting responsibly, though later he acknowledged that early communication with the public could have been better. But he believed that the agency's reaction to the outbreak was largely successful, while granting that it was also a learning exercise. Indeed, there was much to be learned because cases of inhalation anthrax had been so infrequent in the past. Amid the continuing rumble of criticism, Koplan resigned in March 2002.

Despite missteps, which doubtless will be pointed to and debated for years to come, the CDC did provide essential public health support during the crisis. As anthrax was suspected in Florida, then New York City, New Jersey, the Washington, D.C., area, and, finally, Connecticut, CDC professionals arrived quickly at each location. With a few hours' notice, Brad Perkins, Steve Ostroff, Beth Bell, and hundreds of others were on planes out of Atlanta to every hot spot. In cooperation with local officials, they joined in assessing the extent of the disease, providing laboratory analyses, conferring with physicians, and informing the public.

As my conversation with Morse, Kellogg, and Lillibridge wound down, Morse observed, "You know, Julie Gerberding really coordinated the CDC's operations as the events unfolded."

Dr. Julie Gerberding took the escalator down one level from the lobby of the Regency Hyatt and headed right, to Ballroom II. A silver streak threads through her trim dark hair. In a white turtleneck sweater and dark blue jacket, she was a striking figure as she moved to the front of the room. After a few moments, with two fellow panelists, she began to speak on health care issues. Her audience was sparse; there were plenty of empty seats. At the same time, 500 people were packed into Ballroom III next door for the panel on "Anthrax 2001." No surprise—the letter incidents had happened only a few months earlier and were still on people's minds.

Later, Gerberding told me that 2,000 CDC employees had been working full-time on anthrax, "although virtually all of CDC's 8,500 employees participated to some degree." In the fall of 2001 she had been acting deputy director of the National Center for Infectious Diseases, "so I was working here in Atlanta with Dr. Jim Hughes, the director, to try to coordinate the investigation and response." (The NCID is one of 11 component centers and programs of the CDC.)

Articulate and accessible, Gerberding emerged as a principal spokesperson at the CDC's daily press briefings. On November 15, 2001, for example, she fielded dozens of questions during a telephone press conference. Please define "cross-contamination," a reporter from Japan requested. What are the side effects of Cipro? asked the *Atlanta Constitution*. Are the recommended antibiotics safe for women who are breast-feeding? inquired Reuters. Her responses were quick and clear, for which the questioners expressed appreciation.

Gerberding had been at the CDC only since 1998. Before that, she was an infectious disease physician for 17 years at the University of California Medical School in San Francisco. But her managerial skills during the anthrax crisis were so highly regarded that in July 2002, at age 46, she was named the first woman director of the CDC.

Gerberding is at once friendly and authoritative. She praised the leadership of Koplan and Hughes during the anthrax crisis but is unabashedly proud of her own role. "One of the things I was uniquely able to contribute was an immediate appreciation of how important it was to have infectious disease clinical expertise on the ground," she said. She anticipated that wherever anthrax was found, people would show up at medical facilities worried that they had become infected. At the same time, some of them might indeed have the disease. So she was determined to send CDC experts to the field only if they had recent infectious disease experience, "not people who were trained in the remote past."

"Let me tell you another interesting thing from my unique perspective," Gerberding said. She paused, seeming to consider whether she was touting herself unduly, then chuckled and continued:

> When I was in San Francisco, I was very involved in infection control for HIV. And, you know, in the early days of AIDS, people were really frightened about having patients with AIDS in the hospitals. So now we have this patient in Florida who died from anthrax and who is going to have an autopsy. Some people involved with the autopsy were concerned about their own safety. For me, after AIDS, it was just déjà vu all over again.

Dr. Gerberding understood that despite evidence that anthrax was not contagious, people might still fear contacting someone who had the disease. "You don't take away subjective emotion on the basis of data alone," she said. So she sent Dr. Michael Bell, a CDC doctor she had helped train in San Francisco, to work with the autopsy crew in Florida. He was already familiar with fear and infection containment in autopsy suites where AIDS was involved. Dr. Sherif Zaki, the CDC's soft-spoken chief pathologist, was pleased to have Bell come down to Miami with him, recalling that "the pathology people down there were really reluctant to do the autopsy." When Zaki, Bell, and others from CDC arrived, they helped to alleviate the local crew's concerns about risk. Gerberding said, "You know, they realize that if the CDC is willing to do it, it must not be so dangerous."

Basically, Gerberding's role was to coordinate a large number of support systems and field investigations.

> There were so many important scientific decisions that required input from varied areas of expertise—from the laboratory side, from environmental microbiology, from the epidemiology side, the infectious disease side, the communications side—and then to synthesize the input and make decisions about the next step. I was sort of the integrator of information, and with Dr. Koplan, Dr. Hughes, and others would create policies and the next decision steps.

It was Gerberding who helped determine whether a suspected case was actually anthrax, not a simple task when laboratory test results were ambiguous. "So I played the role of calling a case 'a case.'" She emphasized that "a lot rode on whether there was a case or not—you know, closing a postal facility, for example. I didn't take that lightly, believe me."

In dozens of interviews with CDC officials and public health doctors, I asked about their emotional reactions during the anthrax events. Most were stoic. In effect they said, "I'm trained for this kind of thing and my emotions are set aside." Not Dr. Gerberding. "This was a horrible thing," she answered without hesitation. "Of course, I was not dispassionate."

Gerberding's and D.A. Henderson's paths began to intersect more frequently after the anthrax letters. They appeared on panels together and were among the coauthors of a major article in the May 2002 issue of the *Journal of the American Medical Association* titled "Anthrax as a Biological Weapon, 2002." The article was an update of the 1999 *JAMA* article by the 21 experts who wrote about managing an attack with anthrax.

Henderson's earlier career focus had been on eradicating smallpox, Gerberding's on preventing and treating AIDS. In those bygone days, neither had thought a whit about bioterrorism. Now, they were exemplars of a new and pervasive mind-set. Like many other Americans, they had come to view bioterrorism as a threat to the health and security of the nation.

In 1990, Secretary of Health and Human Services Louis Sullivan made a stunning announcement. The United States would seek to map the DNA sequence of the *variola* virus and then "will destroy all remaining virus stocks." Soon after, the Soviets agreed to do the same. The target date for destruction of the two remaining repositories was December 31, 1993. By the end of 1992, scientists at the CDC had sequenced a strain of the virus, but a debate about the wisdom of destruction had already begun.

In his book, *Scourge: The Once and Future Threat of Smallpox* (2001), Jonathan Tucker summarizes the opposing positions. Some supporters of destruction viewed the virus as an absolute evil that should not be preserved. Total extirpation would also, of course, eliminate the chances that the virus might be accidentally released. But even if smallpox were to appear again, people could be immunized by vaccination. (The vaccine is derived from the relatively harmless *vaccinia* virus.) Finally, destroying the *variola* stocks would strengthen the norm against biological weapons. Their elimination would deepen the sense that "it's a crime against humanity to develop such weaponry," according to Jeffrey Almond of the University of Reading.

Opponents of destruction contended that if smallpox reemerged, stored samples of the virus could be useful for comparative investigations. Moreover, the manner in which the *variola* virus caused infection was scientifically interesting and worthy of study. Opponents also felt that the absence of the virus might diminish the sense of need to retain stockpiles of the vaccine. People would therefore be even more vulnerable in the event of an outbreak. As to the moral argument, Wolfgang Joklik of Duke University Medical School retorted that the "symbolism of destroying the remaining stocks of smallpox virus is highly unlikely to influence anyone contemplating biological warfare or terrorism."

During the 1990s, the WHO, through committee decisions and assembly resolutions, repeatedly affirmed the goal of destruction. But debate among scientists continued, and proposed dates for the final act came and went: December 1993, June 1995, June 1996, June 1999. Then in 1999 the World Health Assembly authorized "temporary retention up to no later than 2002 of the existing stocks of *Variola virus* at the two current locations." (The assembly is the decision-making body of the WHO and is made up of health ministers of member countries.) But in 2002, in part prodded by the U.S. administration, the assembly postponed destruction pending further international research. It resolved that the matter be considered again no later than 2005.

Although the initial proposal for destruction came from the U.S. administration, in the mid-1990s opinions among government scientists and medical officials were divided. Those from HHS tended to support destruction while those from the Department of Defense favored retention. D. A. Henderson, who in 1993 had become deputy assistant secretary for health at HHS, had joined in advocacy for destruction at interagency meetings. And he continued his advocacy after leaving the government in 1995. Henderson based his position both on scientific and ethical arguments. He believed that adequate research could be done with preserved *variola* DNA and with other poxviruses (for example, the viruses that cause cowpox or monkeypox.) He also believed that destruction of the declared stocks would make an important moral statement. But at the end of the decade, the scientific underpinnings of his argument came under assault.

In March 1999, at a meeting of the Institute of Medicine, Stephen S. Morse, director of the Center for Public Health Preparedness at Columbia University, was sitting near Henderson. A committee of the institute that had been established to assess "future needs" for the smallpox virus was giving its report. The report concluded that "preserving the live virus may provide important scientific and medical opportunities that would not be available if it were destroyed." A live virus could help to develop new antiviral agents, new vaccines, and new insights into the immune system.

Henderson sat in dejected silence. "It was difficult to watch him at that meeting," Morse said. In effect, the report was saying that destroying the remaining smallpox stocks might *never* be feasible. There would always be a need for new antiviral agents and new insights into immune responses. Years later Morse winced as he recalled his own feelings at that meeting: "It was painful for me to see D. A. in a position where all that ammunition was being presented not to destroy the remaining repositories of the virus."

By early 2003, as the United States was preparing to engage Iraq militarily, worries heightened about Iraq's biological arsenal. Intelligence reports suggested that Saddam Hussein might be harboring secret stocks of smallpox virus. The information accelerated movement toward a new national position on vaccinations. In December 2002, President Bush had directed that vaccinations be administered to 500,000 military personnel and, over time, to 10 million civilian emergency responders. (Twenty-odd people per million who receive vaccinations may suffer serious side effects, and one or two might die. Some people and institutions expressed reluctance to participate in the vaccination program, and by mid-2003 there was talk of scaling back the effort.)

The issue of destroying the stocks of *variola* had been put on hold. Now the more immediate question was whether smallpox, like anthrax, might actually be used as a weapon. The wall that Henderson once thought he had pushed back forever was pushing back at him.

CHAPTER
SEVEN

A SCIENTIST'S RACE TO PROTECTION

It was 9 a.m., and Nancy Connell took the elevator to the basement level of the medical school. Thirty yards down a long quiet corridor she passed a door marked "Caution: Irradiator Room" and another labeled "Stock Room." In the bowels of the University of Medicine and Dentistry of New Jersey in Newark, she turned left at a sign for the Center for the Study of Emerging Pathogens. A card swipe unlocked a door stamped with two red biohazard symbols and a capitalized instruction: Please Make Sure This Door Is Closed Completely Behind You! Fifteen feet farther another swipe opened a second door, to a subterranean suite of offices and laboratories.

Connell, a professor of microbiology at UMDNJ, greeted Paula Trzop, one of her research technicians, at the front desk. Paula handed her a stack of messages, and Connell headed down the hall to her small office. She laid her bulging briefcase on the desk and checked her e-mail. The month was October, one year after the discovery of the anthrax letters, and several list servers were abuzz with stories about the incidents. One reproduced an article from the *Wall Street Journal*, "Armchair Sleuths Seek Anthrax Sender." She read the article and shook her head, dismayed that the perpetrator was still on the loose. Thirty minutes later she ambled across the hallway. Jessica Mann, another technician, was hunched over a laboratory bench, organizing data from the previous day's experiment. "Hi, Nancy, almost done," she said.

Mann was engaged in the Connell lab's most sensitive project. A chirpy 24-year-old, she had been out of college for 3 years and intended to return to study toward a doctorate. But now she was manager of Connell's new high-security laboratory in which experiments with anthrax, plague, and other potential biological warfare agents are conducted. Connell was supervising a 2-year effort to develop a method to quickly determine if someone is infected with any of these bugs. Her work was being funded by $3 million in grants from the U.S. Army Medical Research Institute of Infectious Diseases (USAMRIID). When Connell mentioned the project, her eyebrows rose in wonderment. "I still can hardly believe I'm doing this," she told me.

A few years earlier Connell's study would have been impossible—the research technology she was using did not exist. But neither could she have imagined herself a principal investigator of an Army-backed program. Connell's skepticism of the military was too deep seated. Yet now she was the recipient of grants that supported her work with potential germ weapons, and, to boot, she was named director of the university's Center for BioDefense. The center is receiving an additional $4.5 million from the Pentagon to develop education and training programs and to plan responses to bioterrorism. Brendan McCluskey, the center's deputy director, largely oversees the development of the training and response plans. The head of the hospital's emergency medical services, he holds a graduate degree in microbiology and teaches a course on biological terrorism with Connell. McCluskey, with crew cut and solid frame, presents a rugged appearance. But he acknowledged a profound fear of the tiniest of weapons— germs. "I worry a lot about that stuff," he said. "It's very scary."

Nancy Connell, 49, pondered the road to her current positions. As a youngster she lived with her mother and two sisters in New York City. Guided by her mother's sense of social justice, she went to schools where there was a lot of activism for social causes—first, Saint Hilda's, an Episcopal school on the Upper West Side and then the Oakwood Friends School, a Quaker boarding school in Poughkeepsie. "In those days we were always protesting and marching about something." She thought back to her teens when she joined classmates in demonstrations against the Vietnam War and marched for women's rights, civil rights, and antinukes. For Nancy social justice meant supporting efforts to allocate more money for welfare and less for the military. "When we boycotted our classes," she chuckled, "it didn't mean much anyway because the faculty was boycotting, too."

Connell's social concerns have always been expressed by polite participation, never confrontation. Her brown eyes narrowed when she considered the world's ills, but her demeanor remained calm, influenced by her devotion to things cultural. At Middlebury College, she majored in music and classics. By the time she graduated in 1975, she was an accomplished cellist and could speak and read Latin and Greek.

After college, Connell took some science courses and worked as a research technician and loved doing both. She studied cell biology at Harvard Medical School where in 1989 she earned her Ph.D. "I was a post-doc at the Albert Einstein Medical school when I began to work with a virulent organism, *Mycobacterium tuberculosis*," she said. "That was really exciting." She continued investigating the TB bacterium's genetic activity after joining the faculty at UMDNJ in 1993. Meanwhile, much of her social activism had become channeled into a cause related to her field. "For some time now, I've been working to strengthen the Biological Weapons Convention, the treaty that bans these weapons."

During her years at Harvard, Connell was a member of the Council for Responsible Genetics and chaired its committee against the military use of biological research. She helped gather signatures for petitions, including a pledge by scientists not to engage in research "that will further the development of chemical and biological warfare agents." Her worries about germ weapons drew her to support strengthening the Biological Weapons Convention. "You know, the Convention does not provide for verifying whether a country is cheating," Connell said. In 2002 the Bush administration opposed negotiating a protocol that would allow for international inspections. It believed that inspectors could easily be fooled. "I think that's a mistake," she said. "The right kind of inspections would make cheating much less likely."

Connell also favored strict regulations for U.S. laboratories that handle dangerous microorganisms. In May 1999 she testified before the oversight subcommittee of the House of Representatives Commerce Committee. The hearing was on "Bioterrorism in America: Assessing the Adequacy of the Federal Law Relating to Dangerous Biological Agents."

Congressman Richard Burr of North Carolina asked, "Dr. Connell, you stated in your testimony that academia has not done as good a job as private commercial labs with respect to safety and security. Can you expand on that?" Connell responded, "There has been a lag, but I do think [academic institutions] are catching up." Improvement began, she believes, after universities started receiving fines "for noncompliance in various areas of safety."

The hearing dealt with a central concern of the scientific community and beyond—the tug between unfettered scientific inquiry and the need for security. Ronald Atlas, then president-elect of the American Society for Microbiology, testified that too much regulation could stifle laboratory investigators. A law had been enacted in 1996 that required institutions to register with the Centers for Disease Control and Prevention if they transferred pathogens to other laboratories. That rule, he said, had already inhibited research: "My conversations in the scientific community indicate . . . that a number of individuals are simply not shipping. They are not exchanging."

Connell was unimpressed. When Congressman Burr asked if further restrictions might not discourage research, she responded, "I think it is irrelevant." She made her larger point against the backdrop of the threat of bioterrorism: "I think that a committed principal investigator who wants to work on an organism will work on the organism and will go through the necessary paperwork." She lauded the tradition of scientific exchange but said that inconvenience was warranted because "the world is different now."

Days after Connell appeared before the House committee, she was at a university luncheon seated next to one of UMDNJ's vice presidents, Dr. Lawrence Feldman. "Nancy, I heard you were in Washington last week at a House session on bioterrorism. I was down there, too," he said. "I know. I heard you were testifying to help get some money for the university for a biodefense program," Connell said.

Feldman acknowledged that that had been the goal of his testimony, adding that he had seen a request for proposals from the Department of Defense. "They're looking for research related to biodefense. We ought to be doing more at the university about the threat of biological warfare." Connell asked what he had in mind. "We don't have a handle on how much laboratory work related to biodefense is going on in the university," he responded. "How would you like to organize an effort to find out? And maybe think about applying for a grant yourself?"

Connell was silent. The offer from a top university administrator was flattering, and it seemed to open an opportunity for advancement. Deep inside she had always wanted to make a mark, to be famous. But ambition at what cost? Feldman's proposal was too much to digest at the moment. "Larry, I appreciate your suggestion. I'd like to think about it." When Feldman made the offer, he knew that Connell's interest in bioweapons was to ban them, not to work with them. But given her knowledge of germ warfare policies and her expertise with a respiratory pathogen—TB—he told her she would be the perfect person to spearhead biodefense research at the university. Still, he understood her dilemma. "Sure," he smiled. "Get back to me."

During the next few days, Connell could concentrate on little else. "I kept thinking about the offer. I talked to a few colleagues about it, but mostly I just thought." She also understood that the specter of bioterrorism was growing and that there was a place for science in the fight against it. The words she had recited in Congress the previous week suddenly had new resonance: "The world is different now." Connell believed she could make a contribution. She picked up the phone. "Larry? This is Nancy Connell. I've given your offer a lot of thought. OK, I'll try it."

As a prelude to the establishment of a biodefense center, Connell first addressed the inquiry that Feldman had posed. "I contacted every single department chair in every school at UMDNJ." There were 40 of them.

> I asked for descriptions of any work in their departments that had any relationship to bioweapons, even indirectly. Some were investigating antibiotic resistance, for example. Not of anthrax but of other Gram-positive organisms. That kind of thing. We got a huge number of responses, coordinated them, and wrote this wonderful paper about all the ways that UMDNJ was poised to be involved with biodefense research.

As it happened, plans were under way to construct a special containment laboratory to accommodate Connell and others who were working with tuberculosis bacteria. The facility, known as a biosafety level-3 lab, would be suitable for work on most potential warfare agents as well.

At the same time, along with other scientists expert on genomics and public health, Connell developed a proposal for a 1-year grant titled "Selective Host Transcriptional Response to Virulent Organisms as a Signature Profile of Infection: Application to 'Listed Agents.'" She would be the lead coinvestigator with Jerrold Ellner, a renowned immunologist who was soon to become chair of the Department of Medicine. The "listed agents" were *Bacillus anthracis, Yersinia pestis, Burkholderia mallei, Francisella tularensis,* and *Mycobacterium tuberculosis*. All but the TB organism are considered possible bacterial weapons. The others cause anthrax, plague, glanders, and tularemia.

The aim would be to identify which genes in the host are activated when these bacteria cause infection. A DNA template, or chip, could then be developed to detect each of these infections by sampling blood from individuals thought to be infected. Thus, a fixed record showing which human genes have been turned on could be used as an identification card for anthrax infection. Similarly, distinctive cards could be developed for each of the other bugs. The idea is that by comparing the profiles of activated genes, say, from a sample of blood, it might become possible in hours, or minutes, to determine if the blood has been infected and by which organism. Current techniques can take days before a disease is confirmed.

In March 2000 a letter arrived from USAMRIID informing Ellner and Connell that their proposal had been accepted. Soon after, they submitted a second proposal, this one for a similar study involving four disease-causing viruses: influenza, dengue, hantaan, and monkeypox (which is related to smallpox, though not as dangerous to humans). Approval was quickly granted pending completion of the first study. The grant for the first study had been $1.3 million and for the second $1.7 million. Although the money was now available, the organisms were not. Regulatory red tape and the discovery of anthrax contamination outside the door of a USAMRIID laboratory had put deliveries months behind schedule.

Meanwhile, after the terror incidents in 2001, government backing for programs to thwart bioterrorism mushroomed. Money for bioterrorism-related research at the National Institutes of Health increased fivefold, from about $340 million in 2002 to $1.7 billion in 2003. Anthony Fauci, director of the NIH's Institute of Allergy and Infectious Diseases, oversees most of the new research money. In 2002, NIH dollars were already funding 125 studies under the category of bioterrorism. They ranged from vaccine development to blocking paths of infection. Some, like Nancy Connell's Pentagon-backed project, were aimed at developing faster means of detecting pathogens.

In June 2002 new bioterrorism legislation provided an additional $4.6 billion for increasing stockpiles of vaccines, enhancing security for water systems, and improving hospital preparedness. "Biological weapons are potentially the most dangerous weapons in the world," declared President Bush as he signed the bill. Recalling that "last fall's anthrax attacks were an incredible tragedy," he announced his determination to better "prevent, identify and respond" to bioterrorism. Thus, besides the Department of Defense, other agencies—notably the Departments of Justice, Agriculture, and Health and Human Services—were funding additional projects related to bioterrorism. The establishment of a new Department of Homeland Security at the end of 2002 was intended to help coordinate all these efforts.

The torrent of money prompted some to wonder how efficiently it could all be absorbed. Dr. D.A. Henderson, director of the Office of Public Health Preparedness in the Department of Health and Human Services, had for some years been pleading for more biodefense funds. Now he recited a telling metaphor: "We have been in the desert, praying for rain, and suddenly we are hit with a typhoon. It is indeed overwhelming." Despite a weak national economy, the mood of the nation was supportive. Connell and her fellow investigators were large beneficiaries of the nation's hopes and fears. Would they be able to deliver?

On Tuesday, June 18, 2002, Jessica Mann could hardly contain herself. It was 10 in the morning and the Fed Ex delivery should have arrived. "OK, Jessica," said Nancy Connell. "Let's go see if it's here." The two women walked out of the laboratory suite through the long basement passageway. Two turns and four doors later they were on the loading dock at the rear of the building. Mann approached a clerk whose name tag identified him as James. "Hi. We're here to pick up a package for Nancy Connell's lab. Is it here yet?" Mann asked.

James fumbled through a stack of newly arrived cartons and held one up. It was white, about 2 feet on each side, and not at all heavy. "This one's addressed to Connell," James said. He seemed to take no notice of the biohazard sign on the side or of the bold warning on the label—Infectious Substance: In Case Of Damage Or Leakage Immediately Notify Public Health Authority. The warning was followed by a phone number. The shipper's address was USAMRIID in Fort Detrick, Maryland.

With controlled nonchalance, Jessica showed her identification to James and signed the release form. She cradled the carton and glanced at Connell. They broke into broad grins. With Jessica clutching the package as if it were a newborn baby, they retraced their steps to the laboratory suite. As they entered, Connell exulted to a few of the staff at the door, "Can you believe it? We've been waiting for this day for almost 2 years." Someone held up a glass of water. "A toast," he said.

After their pause for celebration, Connell and Mann headed deeper into the lab suite, took a right turn, and stood before a sealed door. It was the entry to the BSL-3 laboratory built a year earlier to handle highly infectious organisms. The TB bugs that Connell had long been working on could not legally be touched in the more common level-1 or level-2 labs. (Previously, she had traveled to Albert Einstein in New York City to do her research on tuberculosis bacteria.) The BSL-3 lab under the New Jersey medical school was among some 200 in the nation. It cost more than $1 million to build and thousands annually to maintain. (By early 2003 there were four even more elaborate security labs in the United States, plus three additional ones under construction. Designated BSL-4, with their complex pressure and air filtration systems, they each cost about $70 million to construct. Level-4 labs are required for working with lethal organisms for which there is little or no available treatment, such as smallpox and Ebola.)

Connell swiped her card through the panel to the right of the door, just below three red biohazard markers. She entered, followed by Mann who placed the box on a small table. Now in the outer containment area of the level-3 lab, Connell stepped behind a curtain and changed from silk business suit to shorts and T-shirt. Anything heavier would mean more heat and sweat under the protective gear. Mann was already in light clothing. They climbed into jumpsuits made of tyvek, a white plastic material that is impermeable to microorganisms. They zipped the suits up to their throats and pulled on tyvek booties with elastic bands that tightened around the ankles. Then the first layer of latex gloves, which they wrapped with brightly colored masking tape that fastened to the sleeves of the jumpsuits. Next, another pair of gloves, the working pair that must be replaced repeatedly as they become contaminated.

Finally, each woman slipped on a respirator hood. With a clear plastic visor in front, the bottom of the hood extended below the shoulders like a cape. Attached to the back of the hood was an air hose that reached the power pack around the waist. Each breath of inhaled air was forced through a decontamination filter at the top of the hood. With every surface of the body covered by protective gear, they felt the warmth of their suits even while cool air began to blow across their faces.

Connell and Mann glanced at each other's outfits. "Looks good," Nancy said, as she made sure Jessica was fully covered. Mann nodded and signaled a thumbs-up. Hand gestures were an easier means of communication than speech muffled by the masks and the whirring respirators.

Dressed like astronauts about to leave the mothership for a space walk, they stood before the next door and signed in with the date and time. When they opened the door a shrill alarm went off. The air pressure in the inner laboratory was lower to ensure that no germs escaped (if the room sprung a leak, air from outside would rush in, rather than vice versa), and the alarm signaled that the differential in air pressure had been disturbed. They stepped inside and closed the door. Mann reached for the kill switch and the noise stopped. They were in the "common room," facing a table with a computer on it. Batches of sterile test tubes were on the shelf. Supply cabinets lined the walls. Nothing in the lab, including themselves, would leave until it had been washed with a decontaminant.

They moved a few feet to the left in front of yet another door. Connell and Mann each punched in four-digit codes to unlock the door to the "agent room." It was this inner sanctum that would soon house the anthrax and plague bacteria in the sealed container under Mann's arm, germs whose ancestors have killed millions of humans throughout the millennia. Connell thought to herself: "I feel totally excited. We have worked so hard to get to this point—to make this place so good and so safe. But thinking about the power of these bugs to devastate, it haunts my mind."

The space inside was adequate for two people to sit and work or to take a few paces in any direction. On the left were two refrigerators. One contained carbon dioxide, which enhances the growth of certain organisms, including tuberculosis germs. Farther along the wall were a freezer and an incubator. At the far end, opposite the door, stood a centrifuge. The most compelling fixture was the 6-foot-long lab bench on the right, where experiments are conducted. Known as the "hood," it sat behind a glass enclosure. The hood glowed purple from an ultraviolet light that killed germs and kept the area sterile. Mann flicked a switch and the purple was replaced by ordinary white light. "Don't want to kill the anthrax and the plague," she chuckled. They sat on stools in front of the hood, slid the window up, and placed the box inside. Connell narrated as they unpack their "matryoshka," the Russian wooden dolls that contained smaller and smaller dolls in their hollow insides.

> So here's the cardboard box and, inside, a Styrofoam box. Inside the Styrofoam box are cold packs to keep the temperature low, and an extremely sturdy box that is commonly used to transport pathogens. It's very hard plastic, and there's also corrugated plastic around it. Inside that box there's a plastic bag—a sealant bag that contains large test tubes, about 6 inches long. And inside the large tubes are tiny vials the size of my pinkie tip.

In the vials was agar, a gelatinlike material, on which a few streaks of bacteria were visible. "The bacteria look dark brownish," Connell said. "Creamy surface. Smooth," Mann added. "Jessica, I'll get us started, but I want you to handle things too," Connell said, her voice raised to penetrate their head coverings. One by one, Connell lifted four tiny plastic tubes from their chilled surroundings. "Cold enough to keep the bacteria alive but not for them to reproduce," Mann said. Connell nodded. The first vial was labeled "Vollum," a lethal strain of anthrax, named after the British scientist who discovered it in the 1940s. Vollum was a choice germ weapon in the Army's arsenal before the U.S. biological arms program ended in 1969. The second was "Sterne," a relatively harmless strain. Developed in the early 1920s by Max Sterne, a South African veterinarian, it is used to vaccinate animals. The remaining vials contained two strains of plague bacteria, one virulent, the other avirulent.

Connell unscrewed the top from the tube marked "Vollum," inserted a thin plastic loop into the center, and gently stirred. She dipped the moistened loop into a test tube filled with 10 milliliters of a nutrient broth. "Nancy, okay if I do that with the Sterne vial?" Mann asked. "Sure," Connell replied with easy confidence, knowing that Mann had performed similar procedures with other bacteria in the past. The next step was to place the larger tubes into an incubator. Kept at 37° Celsius, the temperature is ideal for bacterial growth. The Fahrenheit equivalent is 98.6°, the temperature of another optimal incubator for bacterial growth, the human body.

The next day, Mann and Connell were back in the lab. The broth was cloudy. That meant the bacteria were growing. Connell smiled and said, "Now, you want to put them in the centrifuge." Mann secured the test tubes in the circular machine—about 3 feet in diameter—and set the dial. The rapid spinning forced the bacteria toward the bottom of the broth. After 5 minutes, Mann lifted the tubes from the centrifuge. "I'll do the first pipette, Jessica, and you do the next." Connell inserted a pipette, a thin plastic tube the size of a large straw, into the broth. She squeezed a lever at the top, which caused a suction action that drew up the liquid. A small clump of claylike material remained at the bottom of the test tube. It was a pellet of anthrax. She covered the pellet with glycerol, a liquid medium that would help sustain the bacteria when they were chilled. Mann duplicated the procedure with the other test tube and then put the tubes in the freezer.

During the following weeks Connell and Mann grew more bacteria and stored them in small vials in the freezer. Along the way, they also confirmed the presence of anthrax by drawing some of the cloudy broth onto an agar plate. After a night in the incubator, the agar was streaked with visible colonies of the bacteria. To the naked eye the colonies looked like lines with creamy smooth surfaces.

After a few weeks, Connell spent less time in the high-security lab, confident in Mann's ability to handle the day-to-day work. Connell needed the time for teaching, meetings, overseeing other research projects, and for directing the university's Center for BioDefense. Since no one may enter the lab alone, Mann was always accompanied by another member of Connell's laboratory staff. Usually it was Paula Trzop, who also has been trained in the special safety requirements of the lab. The first months were devoted to making batches of the bacteria. Mann explained:

> With the bacteria in glycerol they were being preserved in the freezer at –80° Celsius [–112° Fahrenheit]. Pretty cold. We were growing them up to have them ready for the tests with human blood. After a while we had about 10 vials of each strain.

Mann estimated that each vial contained 10 million bacteria. The next step was to convert them from vegetative to spore form. "You just streak the bacteria onto special agar and incubate them at 30° Celsius." That temperature is less than optimal for growth, and reproduction begins to shut down, though the bacteria do not die. Rather, after 7 days, the bacteria have largely transformed into dormant spores.

Meanwhile, in order to obtain blood for the experiments, Connell's staff had posted signs around the medical school complex: "Healthy Human Volunteers Wanted for an Experiment on Infectious Disease." The posters announced that donors who gave blood would be paid $25. A questionnaire filtered out women who were pregnant (whose iron levels could be reduced by loss of blood) and people with AIDS or who were otherwise unsuitable for the experiment. By October, working from a list of eligible volunteers, a physician in the medical school had begun to draw blood. "We take about 240 milliliters from each person," Mann said, "which is about half the size of a soda can."

With a small container in hand, Mann returned to the level-3 lab, where she injected the blood with anthrax spores. Connell was present for the first of the injection sessions. "I knew what to expect, but still the next morning I was shocked," Connell said. During the night the bacteria had split open the blood cells. "It was black. The cells had all been lysed. The devastation was so fast."

Eventually, anthrax and plague genes, as well those of the other agents in the study, would be tested against RNA in human blood. (RNA, ribonucleic acid, is a close structural cousin to DNA.) Only then will it become clear whether the investigation has succeeded, that is, whether the genes that become active are distinct to each organism. The central importance of RNA to the experiment arises from an understanding of its function in gene expression. And this all became possible because of one of the great discoveries of the 20th century.

For more than 100 years, the fundamental characteristics of living things—the shape of a maple leaf, the color of an eye, the infectivity of an anthrax bacterium—have been understood as expressions of microscopic units called genes. Composed of deoxyribose nucleic acid, or DNA, the structure of the gene was identifed in 1953 by James Watson and Francis Crick at Cambridge University. Perhaps the most astonishing aspect of their discovery was the simplicity of the structure. The myriad characteristics of caterpillars, elephants, yeast, ivy, tulips, and amoebas are essentially described by an alphabet of the same four letters. A, T, C, and G comprise the fundamental structures of all genes. They represent four DNA bases—adenine, thymine, cytosine, and guanine.

The bases are joined in matching pairs; "A" is always paired with "T," and "C" is always with "G." The base pairs occur in irregular sequences along a lengthy backbone of sugar and phosphate groups. (The combination of sugar, phosphate, and a base is called a nucleotide.) The structure of DNA resembles a ladder whose sides are made of sugar and phosphates and whose rungs are composed of base pairs.

Splitting the ladder lengthwise results in two separate lines studded with half rungs. Each half rung represents a single base. Thus, if a segment were chopped out from one of the sides, it might contain a sequence of bases that reads TGACCAT. The complementary half rungs on the other side would read ACTGGTA. But a sequence of seven base pairs, as in this example, would make up only a small segment of a gene. Human genes may run as long as several thousand base pairs. It is these lengthy, though uncomplicated, sequences that produce the magic of life.

Watson and Crick immediately understood the profound implications of their discovery despite the restrained language of their published description. Appearing in *Nature*, their article was simply titled "The Molecular Structure of Nucleic Acids" and concluded with a coy observation: "It has not escaped our notice that the specific pairing we have postulated immediately suggests a possible copying mechanism for the genetic material." Only later in his book, *The Double Helix*, did Watson reveal the fullness of their excitement. After the moment of discovery, Watson said that Crick rushed over to a local pub "to tell everyone within hearing distance that we had found the secret of life."

In the years since their discovery, that secret has led to avenues of research that had previously been unimaginable. It has also produced debates about the propriety of the research. In the 1970s the discovery of techniques to insert DNA from one species into another prompted biologist Erwin Chargaff to wonder whether investigators had "the right to counteract, irreversibly, the evolutionary wisdom of millions of years." Such concerns have diminished, however, and gene splicing has become a common research and therapeutic tool. For example, insulin is now produced by bacteria that have been genetically programmed to do so. The human gene responsible for the manufacture of insulin is pasted into the DNA of *Escherichia coli*, a common bacterium that inhabits the human gut. The rapidly dividing bacteria are thus transformed into insulin-producing factories.

Still, the potential for harm from the technology does exist, including in the military sphere. Genes could be spliced into microorganisms that would turn them into super killers that are impervious to any known defense, which is exactly what inspired Soviet activities in the 1970s and 1980s. The sad paradox is that in 1972, the year that the Soviet Union, the United States, and other countries established the treaty to ban germ weapons, the Russians expanded their development program. According to Ken Alibek, a former director of the Soviet program, he and his co-workers ignored the Biological Weapons Convention entirely and assumed the United States was also cheating.

With gene-splicing technology at hand, the Soviets embarked on a program to genetically engineer anthrax, plague, tularemia, smallpox, and other pathogens to make them resistant to antibiotics and vaccines. Only in the mid-1990s, with information provided by Alibek and other defectors, did Americans learn how extensive the Soviet program had been. Until the demise of the regime in 1991, some 60,000 people had been engaged in Soviet germ warfare work.

Meanwhile, in other laboratories throughout the world, advances in legitimate gene research kept apace. The genetic underpinnings of many diseases were identified. Some diseases, such as cystic fibrosis, which can provoke suffocating secretions of mucus in the respiratory tract, are attributable to a single gene. Others, like diabetes and cancer, appear to arise from mutations in several genes.

This knowledge became possible only with an understanding of the structure of the gene. Similarly, the development of research techniques, such as the polymerase chain reaction, depended explicitly on the ability to split strands of the genetic material. Unzipping the DNA ladder is the key to PCR's power to copy a single DNA sequence many million-fold. The new technology also facilitated the cloning of individual genes and the ability to gauge their activity.

In the 1980s a method was developed to anchor a gene segment to a solid surface and stain it with a radioactive or fluorescent dye. The activity of a gene could thus be determined by looking for an interaction between the dyed segment and the target gene from another specimen. Fluorescence changes would signal that the gene was turned on or off.

The actual expression of a gene first involves an unzipped strand of DNA that attracts chemical building blocks to form a complementary strand of RNA. The structure of the RNA is only slightly different from that of DNA, but it has the capacity to direct the formation of a protein. The protein is the end result of gene expression and is what comprises an organism's distinctive characteristics. Thus, if there is active RNA in a blood sample, a gene has been turned on. It is expressing itself.

By the end of the 20th century, efforts to map the entire human genome were nearly complete. The sequencing of every human gene revealed that a person carries perhaps 30,000 genes made up of 3 billion base pairs of DNA. At the same time, a vastly more powerful research capability had been developed, the microarray. Through miniaturization techniques like those used to manufacture computer chips, thousands of DNA probes could be affixed to an area on a glass slide no larger than a postage stamp. Using fluorescent-based detection, the ability to identify gene activity had suddenly been multiplied 20,000-fold.

"I like working with the technology," Amol Amin told me. Amin, who holds a Ph.D. from his native India, has, since mid-2001, worked in Nancy Connell's lab, where he applies his expertise on microarrays. "We are using the oligonucleotide technique to make the chips," Amin explained, using a term that refers to a short chain of nucleotides, or DNA bases. The technique is one of several now being used to make microarrays.

Amin works in the fourth-floor laboratory of UMDNJ's new research building on Norfolk Street in Newark, a half mile from the medical school. The red brick structure, called the International Center for Public Health, was completed in the spring of 2002. Much of Connell's research has been moved there, though not the work dealing directly with biowarfare agents. That is done exclusively at the BSL-3 lab in the basement of the medical school. At the ICPH, Amin handles no pathogens, only the RNA from blood samples that have been infected by the bacteria.

In a small room across the corridor from his laboratory, Amol Amin stood before a 6-foot-wide machine. Labeled "Gene Machine Robot," the apparatus was invented in the late 1990s. The left section of the machine is comprised of several 2-foot-tall metal columns, each containing 10 small circular shelves the size of a computer disk. Sitting on the shelves were plastic plates indented with 384 tiny wells. "Each well contains segments of a human gene immersed in liquid," Amin explained. "And the gene segments are all 70 base pairs long." The gene segments are single stranded DNA to be imprinted on glass slides. The RNA from the blood samples that Mann and Trzop prepared in the BSL-3 lab will later be bathed over the slides to find their complementary DNA imprints.

To the right of the shelves was a flat deck that can accommodate a hundred 1 × 3 inch microscope slides. Dressed in a blue sweater, faded jeans, and sneakers, Amin's latex gloves were the only sign of attire suggestive of a laboratory researcher. He gingerly placed the last of the hundred blank slides on the deck. He turned to the computer behind him and pressed three keys. After a few seconds, the robot began to act. A 10-inch-long red block glided forward from the rear. Its bottom was lined with pins that were spaced precisely to match the tiny wells. The block of pins stopped and hovered over the wells. Like a family of ducks that bob in unison, the pins descended. They came up, paused, and moved laterally, to the right over the glass slides. Again, they descended, leaving imprints on the slides, rose, and retreated to the wells.

Printing the DNA to the slides was but one step in the experiment that Nancy Connell was overseeing. Months of work lay ahead. But 4 months after the project began, Amin and the others on the staff felt a sense of accomplishment. This was the first transfer of genes to slides. These slides will provide the baseline for possible reactions of genes from blood infected with anthrax, plague, and the other organisms. Amin, whose broad smile pierced through a black beard and mustache, intoned, "I have waited for this day." He was about to start working with the first batch of RNA recovered from the blood of volunteers.

The templates that Amin was creating were the core of the experiment. Each of the 20,000 dots on his microarrays was a composite of genes from two samples. "Right now, I am making slides with RNA from blood that is uninfected and from blood that has been infected with Vollum anthrax," Amin said. The slides were washed with fluorescent dye that soaks into the gene segments. If a gene in one sample is active, it will glow red. If active in the other sample, it will glow green. Amin smiled. "But the samples are mixed with each other, and if comparable genes in both samples are active, the gene will glow yellow." He walked over to a computer to demonstrate. "Here are the genes magnified," he said. The computer screen was filled with rows of tightly packed traffic lights—circles of red, green, and yellow. "I will be recording the color of every dot," Amin said. "This will take me quite a long time."

On Wednesday morning, October 16, 2002, Connell was up, as usual, at 5:30, replying to e-mails, reading journal articles, and reviewing notes for the day's lectures. At 7 she went into Eloise's room. "Good morning, sweety. Time to wake up," Connell said, pecking her 9-year-old daughter's cheek. Then it was dressing, frying eggs for breakfast, packing lunch for Eloise. Her husband, Mitchell Gayer, an environmental scientist, was off to work at 7:30. Before 8, Connell walked Eloise a mile to school and then jogged back home. No early meetings that morning, so she had a few minutes to practice her cello. At 8:30 she backed her car out of the driveway. Five minutes later she was on Route 78 west, and in another 20 minutes she pulled into the lot behind the medical school.

When Connell saw Jessica Mann that morning, they discussed Mann's work plan. "Today, mainly I'll be infecting blood samples with anthrax," Mann said, "and checking time points every 4 hours. I guess I'll be in and out of the monkey suit for the next 12 hours." Weeks earlier they had seen the first sample of anthrax-blackened blood. The image remained vivid in Connell's mind as she told Mann to be careful and not to stick herself. Connell reviewed her own schedule. Speaking as much to herself as to Mann, she said she needed to get over to ICPH for class, back to the medical school to give a noon lecture, then to a meeting with some biodefense researchers. A smile from Mann and Connell concluded, "so let's 'talk' by e-mail tonight."

Twenty minutes later Connell was standing next to a table in the middle of a small conference room on the second floor of the ICPH. Her 10 o'clock class was about to begin and she was checking that her Power Point projection was in focus. Six graduate students were squeezed around the table. Another 10 were sitting against the wall. They needed no reminding, but Connell began anyway by saying that they would be talking about *Bacillus anthracis*. "What I'd like to do is start with the capsule and the toxin. Here's a picture of the bacteria under the microscope."

She clicked on the first slide. Clumps of reddish-purple rectangles were everywhere. The rods looked like pieces of All-Bran cereal connected end–to end. "People call them 'boxcars' because they form long chains," Connell said. She clicked on a second slide. The chains were further magnified into curves and twists that were strangely aesthetic. Connell murmured, "They're really lovely," aware of the irony of such a description for something so deadly.

She discussed other bacillus species, including *Bacillus cereus*, a relatively harmless bacterium that anthrax resembles. But the anthrax bacterium possesses a unique cell wall made of a particular sugar. "All virulent *Bacillus anthracis* form this polysaccharide capsule," she continued. Connell explained the basis of the toxin. "What's cool about the anthrax toxin is that it has three factors."

Another slide labeled the factors. One was called "protective antigen," which in effect is the capsule itself. A second was "edema factor," which, as the name suggests, is a protein that induces fluid to accumulate in the area of infection. Third is "lethal factor," another protein.

Distinctive genes are responsible for the synthesis of each factor. But the genetic expression of all three is necessary for the toxic effect. "You need all three toxins to cause lethality of a macrophage," Connell said, referring to the body's cell that ordinarily engulfs and destroys a foreign invader.

She inserted a comment about anthrax as a bioweapon. "So it is the toxin that kills. We don't care about the bacteria, just the toxin." She noted that biological agents on the skin are unlikely to cause problems unless they land on a scratch or other opening. Yet during some recent anthrax threats, people were made to undergo vigorous scrubbing. "That is the wrong thing to do because you can abrade and open the skin," Connell said.

Dr. Connell spoke quickly, enthusiastically. She paused intermittently to allow her listeners a moment to absorb her point. The result was speech delivered in packets bracketed by cushions of silence. Her voice, though soft, sounded eager, as if coming from someone on the edge of a chair. Connell's style seemed just what her students appreciate. As a tribute, she has repeatedly won the "Golden Apple," an award designated by the school's graduate students for excellent teaching.

A student in sandals who was sitting across the table raised his hand. He said he had read that anthrax growth is enhanced by contact with carbon dioxide. "Does that mean that the effect of inhaled anthrax is made worse because of carbon dioxide in the lungs?" he asked. Connell mulls over the question, one that she had not thought about before. "I guess there might be pockets of carbon dioxide where that could happen," she said, acknowledging uncertainty. She said she'd research the question and have an answer for him next week.

During the last minutes of the class, Connell asked if everyone had read the paper she had assigned. Heads nodded in the affirmative. "So let's talk about it," Connell said. Titled "A Bacteriolytic Agent that Detects and Kills *Bacillus anthracis*," it had appeared two months earlier in *Nature* and described a study led by Vincent Fischetti of Rockefeller University in New York City. "Who can tell me what a phage is?" Connell asked. Hands shot up, and Connell pointed to a young woman in a brown sweater. Her answer, delivered with a slight Spanish accent, was brief and on the mark: "It's a virus that infects bacteria." "Exactly," Connell said. "The 'bacteriolytic agent' referred to in the title is an enzyme of a particular virus called gamma phage virus."

Mention of gamma phage harks back to the laboratory test that Phil Lee had performed the previous year to confirm the presence of anthrax in Bob Stevens. Now, Fischetti had isolated the active ingredient of that virus. The article described how the ingredient kills the anthrax bacillus by "lysing" it. "Lysis," Connell explained, "means that the bacterial cells are blowing apart and everything is leaking out." Fischetti and his coauthors concluded that the isolated ingredient could also be "exploited as a rapid method for the identification of *B. anthracis*." The sentence resonated for Connell. Although she did not say so in class, the central effort of her own experiments with biowarfare agents is to develop a method of rapid identification.

At 11:40 Connell ended the class, apologizing that she could not linger for questions. She gathered her notes and computer into her briefcase. It was raining, so she would be driving the half mile back to the medical school for her noon lecture there. She donned a dark gray raincoat and plunked an old brown fedora on her head. "It was my grandfather's. I took it after he died in 1978, and I wear it in the rain," she told me. Her sartorial selection seemed at once a gesture of sentimentality and a carryover of 1960s nonconformity. As she scurried across the street to the parking lot, Connell's seamless movements appeared more like the glide of a hydroplane than a sequence of footsteps.

In the ground-floor kiosk at the medical school, Connell purchased a dollar bottle of spring water, and rushed up a dozen steps to lecture hall 610. She would speak from a pit, looking upward toward seven ascending rows of seats. Coarse cream-colored bricks lined the walls. The lecture was open to the medical faculty, and a dozen men and women in white coats were already seated. Another 20 trickled in as Connell prepared her Power Point. "Today I'm going to talk about two things—some of my activities with TB in the lab and then a bit on the Center for BioDefense." Again, Connell spoke in enthusiastic spurts. Her curly brown hair, an inch short of her shoulder, bobbed when she turned to the screen and back to her audience. "So here is a graph showing the time that it takes for the metabolism of arginine within the mycobacterium cell."

After 45 minutes, Connell took a few questions and then briefly described the Center for BioDefense. "It was established in 1999, as many of you know, and we are continuing to get great support." She summarized a few of the center's projects, including her own, and took the last question at 1:10 p.m. From there she moved quickly, one flight down the escalator, and a right turn. A few hundred feet and Connell entered the cafeteria. Ten minutes late, she joined nine people in a back room off the cafeteria for a luncheon meeting. They were research scientists, all associated with the Center for BioDefense. The scientists represented a range of specialties—David Alland, thin with a young face, is involved in tuberculosis diagnostics. Elizabeth Raveche, tall and blond, studies laboratory medicine with animal models.

Connell apologized for being late. David Perlin, who heads the Public Health Research Institute, which is associated with UMDNJ, answered, "That's okay, Nancy. We wouldn't start without you." He smiled, joined by a few others, knowing that Connell was supposed to chair the discussion. "Let's order the food first," a voice said. A waitress announced that the grilled chicken and vegetables were especially good. "I'll have that," Connell said, "and a Diet Coke." Each person placed his order.

"So what projects do we want to put forward?" Connell asked. The National Institutes of Health is providing $10 million for research institutions in the New York-New Jersey area, she said. UMDNJ and the Center for BioDefense would like to get a bite of at least a half million. Someone suggested, "We'd want to start by infecting small animals with select bioagents, work up to larger animals, then to natural hosts." To which Raveche responded, "Well, I think we ought to consider transporting human cells into mice. Diagnostic and vaccine testing then could be done in one model."

As the discussion continued, Perlin suddenly got everyone's interest. "We ought to be trying to build a BSL-4 laboratory." He said the first few million could be available from current funds, but for additional funding "we need political support." It was clear that he had been thinking about the matter. He elaborated:

> Cliff Lacy, the state commissioner of health, has federal money to build a BSL-4, and there is a push by some to build it in New Brunswick. But the mayor of Newark will support us, and we have to show there is community support. It's a matter of leadership.

A few people mentioned contacts they had who might be helpful and the need to get one of the university's vice presidents behind the effort. Perlin created a large vision:

> We are really talking about a comprehensive center for biodefense—with BSL-3 and BSL-4 labs, other labs, offices. The New Jersey Medical School would have to take this on as a priority. An NIH grant even of $10 million would not be enough. More would probably be coming from the new federal Homeland Security Department.

As the discussion continued, people at the table expressed enthusiasm about Perlin's ideas. No one questioned whether the university should be seeking a BSL-4 lab but just how it could be done. Still, scientists elsewhere had begun to voice concerns about the growing number of laboratories equipped to do work with warfare agents. Richard Ebright, a Rutgers biochemist, worried about level-3 as well as level-4 laboratories. "It is difficult to conceive of scenarios," he said, "under which increasing the number of persons with access to, and training with, agents such as *Bacillus anthracis*, *Yersinia pestis*, and *Francisella tularensis* would enhance—rather than degrade—national security."

Nancy Connell had been quiet during the discussion about seeking a BSL-4 lab. Months earlier she had signed a letter with Ebright to *Nature* magazine that favored limiting the number of institutions with access to bioweapons agents. "I'm aware of Ebright's concerns," she told me after. "We still have to figure out the right balance." Connell filled the rest of the day with a visit to her postdoc and two graduate students who were conducting tuberculosis experiments. They were at the laboratory at the ICPH, where they had access to the new BSL-3 facility just completed there. She thought about Perlin and Ebright, both decent people but with such different perspectives. She contemplated how far she had come in her own thinking.

> You know what? I still sometimes think, "God! What am I doing?" But my opinion of people in the Army now is completely different. You know, a lot of them are wonderful. I mean they're interested in science, committed to figuring out how these organisms work.

She burst into laughter. "Who'd ever have thought that my best friends would be lieutenant colonels?"

Nancy Connell, like many others in science and in law enforcement, intelligence, and public health, is in a contest against bioterrorism. They are racing to protect Americans from the kind of anthrax horror the country experienced in 2001. The result of Connell's research, like that of hundreds of other current bioterrorism-related science projects, may not be known for some time. But the fact of her engagement reflects the altered mind-set shared across the country. Connell still feels the tug of her social activist roots. But she is no less drawn to the challenges peculiar to these times.

The challenge presented by the anthrax letters was novel in many ways, though not all. The letters were not the first anthrax threats by mail. Prior to the actual anthrax incidents, there had been hundreds of hoaxes. Commonly, a letter would contain powder and a message that the reader had been exposed to anthrax. While all those threats proved to be false alarms, for some recipients the experience had been terrifying.

CHAPTER
EIGHT

TERROR BY HOAX

Clayton Lee Waagner, 45, stared straight ahead. His brown shaggy hair touched the back of his collar and a mustache drooped over expressionless lips. It was this mug shot that had adorned FBI and U.S. Marshall Service posters of the Most Wanted. Then on December 5, 2001, just before 1 p.m., Waagner was apprehended at the Kinko's copy center on Princeton Pike in Springdale, Ohio, 10 miles north of Cincinnati. An employee recognized his picture and called the police. Waagner reportedly was trying to program a computer to fax threats to abortion providers around the country. "We knew he frequented Kinko's," said Deputy U.S. Marshall Bruce Harmening. "We'd been watching the Kinko's and we had a flier in every one in America."

Ten months earlier, in February, Waagner had escaped through the roof of the county jail in Clinton, Illinois, where he was awaiting sentencing on firearms violations and automobile theft. Now, with his recapture 2 months after the anthrax crisis began, he faced additional charges. The arresting officers had found a bag of white powder in his stolen Mercedes. Waagner told them the powder was not anthrax but that initial tests might suggest that it was. A February 2002 publication of the Feminist Majority Foundation reported that the powder "initially tested positive for anthrax [but] later turned out to be an insecticide called *Bacillus thuringiensis*."

The day after Thanksgiving 2001, Waagner visited Neal Horsley at his home in Carrollton, Georgia. Horsley, an antiabortion militant, taped their conversation during which Waagner read off a list of Federal Express billing numbers. He recited the numbers to prove that he had sent hundreds of threat letters to abortion clinics in the first week of November. They were in addition to another batch of threat letters he had sent via the U.S. mail in mid-October. Altogether Waagner was charged by federal authorities with mailing about 550 letters to abortion facilities in 24 states with the intention of shutting them down.

At court hearings in early 2002, Waagner was sentenced to 49 years on charges related to his escape and to theft and weapons violations. But in a Philadelphia court on October 17, 2002, he pled not guilty to the new federal charges involving the threat letters. Despite his taped admission, Waagner denied threatening to use a weapon of mass destruction and making and mailing threatening communications. The indictment cited a message that Waagner had posted in June 2001 on a Web site called "ArmyofGod.com." The message read in part:

> The government of the most powerful country in the world considers me a terrorist. . . . They're right. I am a terrorist. To be sure, I'm a terrorist to a very narrow group of people, but a terrorist just the same. As a terrorist to the abortionist, what I need to do is evoke terror. Thus the reason for this letter. I wish to warn them that I'm coming.

Waagner warned that he "would go after" anyone who worked at an abortion facility and that "I'll drop you a note and we'll get this terrorism thing started in earnest." The indictment also indicated that "on or about October 12, 2001," Waagner mailed numerous letters to reproductive health care clinics. The envelopes contained powder and a letter saying that "you have been exposed to extremely high levels of *Bacillus anthracis* (commonly known as anthrax)." The letter explained further:

> We are a small cell of the Army of God known as the Virginia Dare Cell. After many years of taking the passive course, we have come to the difficult position of having to do that which is right. We cannot defend the pre-born child in the Senate, nor in the Courts. You've won those battles. This letter is to put you on notice of a new era in the battle to protect the pre-born child. We will fight you on the streets of America. We will fight you from the shadows. We destroy your houses of death with fire and explosives and we will destroy you with the three B's (bullets, bombs, bio-weapons).

The letter concluded with "Stop now or die" and was signed by the Army of God—Virginia Dare Cell.

The batch of letters sent "on or about November 7, 2001" by Federal Express said: "You ignored our warnings, so now you pay. Enclosed you'll find the real thing—Anthrax, very high grade. Be careful not to bring any of the spores home to your children." Again, the sender was the Army of God—Virginia Dare Cell.

Before the powder was found to be harmless, according to the indictment, people at the targeted facilities underwent decontamination procedures and unspecified medical treatment. At the court hearing in October 2002, Waagner acted as his own lawyer. He would continue to do so, he said, although he had accepted legal counsel from the public defender's office in Philadelphia. Pending a future trial, he remains at the federal penitentiary in Lewisburg, Pennsylvania, 110 miles northwest of Philadelphia. Waagner, the most productive of bioterrorism hoaxers, had engaged in a phenomenon with a surprisingly brief history.

One block from the National Geographic Society in Washington, D.C., across the street from the Governor's House Hotel, an eight-story building stands at 1640 Rhode Island Avenue. It had been the headquarters of B'nai B'rith, a Jewish service organization, until 2002 when the organization moved to another Washington location.

At 7:45 a.m. on April 24, 1997, Carmen Fontana, the chief of security for the B'nai B'rith building, pulled his pickup truck into the building's underground garage. He took the elevator to his small office on the second floor, picked up his messages, and went down to the security desk in the lobby. Dressed in a dark blue blazer and tie, he greeted the 100-odd employees as they filed past him during the next hour.

"Good morning, Mr. Berk," Fontana intoned in a warm gritty voice. Just the right sound for a former police officer, Harvey Berk thought. Fontana had retired from the D.C. force 6 years earlier and then came to work at B'nai B'rith. "Morning, Carmen. Supposed to be a beautiful day," Berk answered. The sky was cloudless and the temperature rose to 60° by midday. Berk, director of communications for the organization, headed to his office in the 7th-floor executive suite.

During his years with the organization, Fontana had seen plenty of crackpot mail. But the one that Rusty Mason, the mail clerk, showed him that morning was different. At 11 a.m., Rusty came down from the second floor mailroom with a stuffed 8 × 10 inch manila envelope. Lined with bubble wrap, the envelope was stapled along the sides. "It was leaking this red substance," Fontana said. "We were thinking, 'bomb.'" He and Rusty placed the package in a waste paper basket and ran outside with it. Fontana then went back and dialed 911. Minutes later the police arrived.

The bomb squad took an X ray and found no evidence of an explosive. But inside the package they came upon something unexpected. A broken petri dish was oozing a red gelatinous material. The dish was marked "Anthracis Yersinia," according to an after-action report prepared by the Federal Emergency Management Agency (FEMA). (Some news reports indicated that the dish was labeled "anthrachs.") A threat note was also in the package, but police would not reveal its contents other than to say it was written by someone claiming to be with the "Counter Holocaust Lobbyists of Hillel."

Out of concern that the package contained anthrax, the fire department's hazardous materials (HAZMAT) unit was summoned. The unit isolated a number of people who had been close to the package. Fire officials sealed off a one-block area around the building. The 109 people inside were quarantined, as were unknown numbers in nearby businesses, offices, and the Governor's House Hotel across the street. Behind a graying mustache, Fontana offered a sense of mature calm. But his insides were churning: "I was a little nervous when I thought it was a bomb. When I found out it was anthrax and [realized] I had taken a deep breath of it, I believed I was dead," he told me.

HAZMAT and emergency medical personnel consulted with officials from the Centers for Disease Control and Prevention, who advised that the victims be decontaminated with a 1 percent bleach solution. Meanwhile, the package had been turned over to the FBI and then delivered to the National Naval Medical Center in Bethesda, Maryland. The tension and confusion were captured in the FEMA report:

> Personnel then waited for the results of the testing. During this time, a security guard in the quarantine area developed chest pains. He was carried on a chair through the decontaminated corridor and then transported to a local hospital. . . . Also during this waiting period, several MPD [Metropolitan Police Department] officers became up-

> set with instruction that they undergo decontamination. The officers had become aware that the media [were] broadcasting live pictures from cameras positioned on top of a nearby building. The officers refused to disrobe and undergo decontamination. One of the officers struck the EMS [Emergency Medical Services] lieutenant assigned to the quarantine area. High-ranking police officials were asked to help get the officers to comply with the procedures and, eventually, the officers were decontaminated.

The quarantine ended after 9 hours, when laboratory analysis confirmed the material in the package was not dangerous.

The FEMA report highlighted the shortcomings of the response procedures. Fontana's summary is more direct. "No one really knew what to do," he told me afterward. He said that the responders went to a supermarket and bought Clorox bleach off the shelf. They then sprayed it at the 30 civilians, police, and fire personnel who had stripped to their underwear. After decontamination, the victims were given coverall suits and escorted to a bus. "We sat on the bus for 8 hours, reeking of Clorox," Fontana said.

He elaborated on his initial reluctance to obey the directive to stand in front of the building, undress, and be sprayed. "I got into an argument with the fire commander. I told him I did not want my 14-year-old granddaughter to see me stripped down on television." The fire commander said that the decontamination site would be next to a fire engine, which would block the view of outsiders. However, television cameras on a nearby rooftop had perfect visual access. Videos of the event, including the spraying, appeared on the evening news. "Ironically, my granddaughter did see me on TV, stripped. She panicked," Fontana said. Another B'nai B'rith employee was so embarrassed by the ordeal that he never came back to work again.

The FEMA report noted that the absence of tents for decontamination meant that people had to "disrobe in front of television cameras." As a result of the experience, the report hoped that HAZMAT departments in the future would consider "the public's modesty." The report also cited contradictory views about whether people should have been quarantined in the first place. It referred to unnamed "experts" who said that by "isolating people in an unventilated and possibly contaminated area, the victims were in effect exposed . . . for an extended duration." The report's concluding observation left the matter in limbo: "More research is needed to address the question of whether it is best to protect in-place or to evacuate to a safe haven, those civilians exposed to chem-bio agents."

It was assessments such as these, that failed to even distinguish between "chem" and "bio" agents that distressed people like D.A. Henderson. The Keystone Kops response to the nation's first ostensible anthrax threat made clear how ill prepared the authorities were. Moreover, few people realized that the hoax attack on B'nai B'rith was just the beginning of a new kind of epidemic.

In the 18 months after the B'nai B'rith incident, only a few other anthrax hoaxes were reported, though the frequency began to accelerate toward the end of 1998. The incidents usually involved letters claiming that the reader had been exposed to anthrax or telephone calls claiming that bacteria were in the ventilation systems. A report by the CDC in February 1999 reviewed seven recent bioterrorism threats. Responses continued to be chaotic and inappropriate. At some locations, presumed victims were told to go home, place their clothes in plastic bags, and shower. At others they were quarantined, made to disrobe, to undergo scrubbing with diluted bleach, and to begin taking antibiotics. In a few cases, victims were hospitalized.

These responses reflected contradictory guidelines from different agencies. A 1998 anthrax advisory by the FBI said that if "there is confirmation" of anthrax exposure, victims should undergo decontamination with a diluted household bleach solution, specifically "Clorox—5.25 percent hypochlorite." When confronted with a threat, many local authorities ignored the requirement to confirm the presence of anthrax. Erring on the side of caution, they required potential victims to be scrubbed anyway. The CDC's February report advised that presumed victims be washed with soap and water but also suggested decontaminating "the environment in direct contact with [a possibly contaminated] letter or its contents . . . with a 0.5 percent hypochlorite solution (i.e., one part household bleach to 10 parts water)." Whether or not the "environment" included the people who had been in direct contact was not made clear.

A June 1999 document prepared by the Association for Professionals in Infection Control and Epidemiology (APIC) in cooperation with the CDC was more definitive: "Decontamination should only be considered in instances of gross contamination" In bold print the document also stated: "Potentially harmful practices, such as bathing patients with bleach solutions, are unnecessary and should be avoided." The APIC statement apparently was the first to explicitly warn against washing people with bleach because it could cause them harm. Moreover, it was followed in August with an observation by CDC officials that "the actual efficacy of disinfection with sodium hypochlorite [bleach] in cases of bioterrorism is questionable."

Scrubbing victims with bleach solutions was not the only "questionable" practice of emergency responders to bioterrorism threats. Many incidents involved other erratic actions as well. Inconsistencies may be seen in Table 1, which lists information on 40 anthrax hoaxes. Representing about 20 percent of the total of some 200 incidents that occurred between April 1997 and June 1999, they were selected because of either of two criteria: first, available information indicated large numbers of people were affected or, second, aggressive treatments were rendered. The data were drawn from news articles, official reports, and personal interviews with officials and victims.

Actual numbers of evacuated or quarantined victims were available for only 30 of the 40 listed incidents. These totaled 12,398. The numbers reported for the other 10 listed incidents were variously described as "several" or "hundreds." Thus, for the 40 incidents, the number of lives disrupted approached 13,000. For the approximately 200 incidents, the figure surely exceeded 13,000.

Among the 40 listed hoaxes, in 26 of them, the victims underwent decontamination (usually with bleach solution), and in 20 they received antibiotic treatment and/or were hospitalized. In 30 of the incidents (75 percent of the total), potential victims were subjected to at least one of these aggressive responses.

Table 1 reveals the breadth of inconsistency but nothing about the psychological effects on the victims. Interviews indicated that several people in the targeted areas were unfazed and considered the episodes minor inconveniences. But many were terrified. Similarly, some felt that emergency response teams acted efficiently and sensitively. Others, even months after the experience, were bitter about having undergone embarrassing and intrusive treatment.

Thus, 2 years after the B'nai B'rith event, local response teams were still repeating the errors of that experience. Responsibility largely lay with the delayed introduction of clear and rational protocols by the CDC and other national agencies. Descriptions of several incidents demonstrated the variety of responses and their effects on the victims.

The NBC network news office in Atlanta, Georgia, is on the 11th floor of a 12-story office building. Around noon on Thursday, February 4, 1999, an employee opened a mailed envelope containing a baggie full of material that looked like dark sand and pepper. An accompanying letter said: "You and everyone in this building have been exposed to anthrax." People in the office called 911, the FBI, and the CDC. Two plainclothes police officers quickly came over from the police headquarters across the street. Later, more emergency responders arrived, including additional police and paramedics.

All seven NBC employees and the two plainclothes officers were sprayed with a bleach solution while in the office and fully dressed. Then, in wet clothing the nine victims were taken to the elevator and down to the lobby. The lobby was filling with people from other floors who had learned of the anthrax threat and were trying to leave the building. As the nine victims got off the elevator, other police officers ordered them to return upstairs. Meanwhile, officials were sealing off a four-block area. Some of the 600 people in the area had been evacuated but most remained under quarantine.

Fire department personnel arrived 45 minutes after the first telephone calls. Some were dressed in "space suits" and went up to the 11th floor, where they sprayed the hallways and the NBC office with a bleach solution. The nine victims also received a second spraying, again while fully clothed.

An NBC staff member recounted her experience to me. Felice (a pseudonym) was among the nine who were sprayed. Her clothing, like that of the others, had turned pale from the bleach. After the second spraying, the nine were again taken to the lobby by elevator. This time they were escorted out of the building to awaiting ambulances. Still clothed and drenched, Felice and another woman got into one of the ambulances but were immediately ordered out because "someone said we might be contaminating the ambulance."

The two women were escorted to a nearby HAZMAT truck where, along with the other victims, they were decontaminated again. One at a time the nine victims undressed and submitted to a third bleach washdown. Following a shower in the truck with plain water, they were given sheets to wrap themselves in. As they exited the truck, a man held a tarp to block the view of onlookers. There were not enough ambulances for everyone, so the five women were taken first and the four men were left to wait until other ambu-

TABLE 1 Forty Selected Anthrax Hoaxes, April 1997–June 1999 (based on whether victims were reported to have received treatment or if large numbers were affected)

Date	Location	Target	Threat Communication
6-15-99	Hempstead, NY	Courthouse	Letter
6-15-99	Mineola, NY	Courthouse	Letter
5-17-99	Syracuse, NY	College (Catholic)	Letter
4-22-99	Kansas City, MO	IRS offices	Letter*
4-19-99	Biloxi, MS	Courthouse	Note
3-3-99	Salt Lake City, UT	Church	Package and note
2-26-99	Boise, ID	Planned Parenthood	Letter
2-24-99	Salt Lake City, UT	Planned Parenthood	Package and note
2-23-99	Pittsburgh, PA	Abortion clinic	Letter
2-22-99	St. Louis, MO	Planned Parenthood	Letter
2-22-99	Kansas City, MO	Planned Parenthood	Letter
2-22-99	New York, NY	Planned Parenthood	Letter
2-18-99	Milwaukee, WI	Womens Health Organization	Package and letter
2-18-99	Cincinnati, OH	Abortion clinic	Package and letter
2-18-99	Manchester, NH	Planned Parenthood	Package and letter
2-18-99	Spokane County, OR	Planned Parenthood	Letter
2-12-99	Los Angeles, CA	*LA Times* Building	Letter
2-4-99	Atlanta, GA	NBC news office	Package and letter
2-4-99	Washington, DC	*Washington Post*	Letter
2-2-99	Lackawanna, NY	High school and library	Phone
1-13-99	Tualatin, OR	Library and city offices	Phone
1-2-99	Anaheim, CA	High school	Phone
12-31-98	Rochester, NY	Federal building	Letter
12-26-98	Pomona, CA	Nightclub	Phone
12-24-98	Palm Desert, CA	Department store	Phone
12-23-98	Chatsworth, CA	Cable TV company	Phone
12-21-98	Van Nuys, CA	Two courthouses	Phone
12-18-98	Woodland Hills, CA	Courthouse	Phone
12-17-98	Westwood, CA	Office building	Letter
12-4-98	Coppell, TX	Post office	Vial*
11-18-98	Miami Beach, FL	Fashion magazine office	Letter
11-10-98	Seymour, IN	Wal-mart and bank	Letter
11-10-98	Bloomington, IN	High school	Letter
11-9-98	Indianapolis, IN	Church and school (Catholic)	Letter
11-9-98	Cheektogawa, NY	Church (Catholic)	Letter
11-2-98	Toledo, OH	Abortion clinic	Letter

Threat Material or Technique	Evacuated or Quarantined (approx.)	Decontaminated (some or all)	Hospital and/or Antibiotic (some or all)
Brown stain	500		
Brown stain	500		
White powder	30	Yes	Yes
Dark substance	2,000	Yes	
Ventilation system	100		
Contents of package	Several	Yes	Yes
Powder	24	Yes	Yes
Contents of package	20	Yes	Yes
Contents	3+	Yes	
Contents	Several	Yes	
Dark smudge	27	Yes	Yes
Contents	Several		
Smudge	Hundreds	Yes	Yes
Contents	4	Yes	Yes
White powder	Several	Yes	
Contents	3+	Yes	
Gray powder	2,000		
Black powder	600	Yes	Yes
Contents	Several	Yes	
Spread in buildings	Several		
Spread in buildings	Several		
Ventilation system	120		
Contents	1+	Yes	Yes
Ventilation system	800		
Ventilation system	200	Yes	
Unspecified	200	Yes	
Ventilation system	2,000		Yes
Ventilation system	90	Yes	Yes
Contents	21	Yes	Yes
Contents	500	Yes	
White powder	19	Yes	Yes
Contents	Several		Yes
White crystalline material	1,800		
Brown powder	480	Yes	Yes
Contents	9	Yes	Yes
Contents	3	Yes	Yes

TABLE 1 Continued

Date	Location	Target	Threat Communication
10-31-98	Louisville, KY	Abortion clinic	Letter
10-30-98	Indianapolis, IN	Planned Parenthood	Letter
8-18-98	Wichita, KS	State office building	Note
4-24-97	Washington, DC	B'nai B'rith	Mailed petri dish

*Did not name contents but assumed to be anthrax.

NOTE: The information in this table is based on news reports and interviews with officials and participants in the incidents. The 40 listed incidents represent about one-fifth of the total number of anthrax hoaxes recorded by the FBI during this period.

lances arrived. Once in the ambulance, Felice was told she would have to be hooked to an intravenous (IV) line. She objected strenuously. "I hate needles," she said, but was told she had no choice.

Upon arriving at Grady Hospital, the victims found that outdoor showers had been set up on a platform in front of the entrance. Felice and the other women were told they could not enter the hospital until taking a shower (their fourth decontamination). "The area was not closed off, and we had to take off our sheets and go under the shower in public view," she told me. While still attached to the IV line, Felice washed herself with the detergent soap she was given. "For some of us, the water was so hot it felt scalding. But others had very cold water. It really was erratic."

Inside the hospital the five women were given hospital greens to wear. Soon after, they were told that initial tests showed no evidence of anthrax. The test results became available before the male victims were brought to the hospital so they were spared the shower in public view.

Felice said she felt less troubled by the experience than some of her co-workers. "They were angry they had to go through this,"

Threat Material or Technique	Evacuated or Quarantined (approx.)	Decontaminated (some or all)	Hospital and/or Antibiotic (some or all)
Brown powder	Several		Yes
Brown powder	31	Yes	Yes
White powder on stairs and elevators	200		
Red gelatinous substance	109+	Yes	Yes
Totals:	12,398 (+10 incidents reporting "several" or "hundreds")	26	20

she said. "The thing that mainly got to me personally was having to have the IV. For what purpose?" In the days afterward, Felice thought about the event and how poorly it was handled. She wondered why the men holding the tarps up were not in "moon suits" when all the other personnel were. "That made no sense to me." Also, she said that it had been difficult to understand what people were saying while in their masks. Much of the time the victims tried to interpret hand motions by the personnel in masks because their verbal commands were unintelligible. Moreover, Felice felt frustrated by a lack of singular direction. "There were a lot of different agencies, but it was not clear who was in charge."

The day after the incident the victims were visited briefly by a group of health officials from the CDC. The officials told them that the chances that anyone was exposed to anthrax were very remote but that they would not know for sure until laboratory tests were completed in the next few days. Later, when learning more about anthrax, Felice wondered, "In that case, shouldn't we have been taking antibiotics?"

Felice thought that the CDC and other agencies might have

gained knowledge by inviting the victims to discuss their experiences at length. Hearing from the victims could help responders understand the psychological dimension and improve their actions in the future. Four months after the event, when I spoke to her, Felice was still lamenting the lack of such follow-up. Although the incident was no longer a central concern at the NBC office, "we all watch the mail a little more cautiously now," she said.

Meanwhile, according to Lieutenant Reggie Latimer, who led the fire department's response at the Atlanta NBC site, the experience taught everyone a lot. In the future he said, "we would not be so quick to have people decontaminated under those circumstances."

On Monday, February 22, 1999, around 9:30 a.m., as a snowstorm raged outside, an employee of the Planned Parenthood clinic in midtown Kansas City, Missouri, was opening the mail. She came upon a letter containing brown powder, a skull and crossbones, and a statement that the reader had just been exposed to anthrax. A call to local authorities set in motion a municipal response led by the fire department's HAZMAT team, police, FBI, and medical personnel. Before the call was made, a staff member (who asked not to be identified by name) had seen the letter and gone home: "When I saw the dirty-looking piece of paper with a skull and cross bones and the word 'anthrax,' I just left. I did not want to be around," she told me. At home with her 3-year-old son, she began to think: "What if I'm infected?" She became increasingly worried and called the clinic. A firefighter told her to take a bath with bleach and water and that someone would call her if there were a problem. About 5 hours later, she learned from a television broadcast that authorities determined that anthrax apparently was not present. But she remained concerned because "we were not allowed back in the building for two days."

When firefighters first arrived at the clinic, they quarantined 20 people who were there and then set up a tent for decontamination outdoors. The detained people—five men and 15 women—were told they would have to undergo a washdown with diluted bleach. Seven firemen who had entered the building were also ordered to undergo decontamination even though they had been wearing masks and protective outerwear. The women went through the process first while the 12 men waited their turns in the building.

One at a time the victims went outside in the freezing weather. A few men held up tarps to block the view as each victim undressed. Then, completely naked, each victim entered the unheated tent. Once inside, the victim stood in a large basin of water as a woman in a mask and protective gear scrubbed him or her with a bleach and water mixture. Following the scrubbing, the victim was hosed off with cold water. Warm water was not available, according to Chris Bosch, division chief of research/planning and emerging technologies, Kansas City Fire Department. After decontamination, victims were given disposable blankets to wrap themselves in and then exited the tent. They walked barefoot to a waiting vehicle. The civilians were brought to a city bus parked nearby, and the decontaminated firemen were brought to a HAZMAT truck. Once aboard, they dressed in dry wrappings, sweatsuits, and shoes provided by the Salvation Army.

"We were all embarrassed to strip," said Greg Ono, one of the seven firefighters who went through decontamination, "but some objected more than others." He felt sympathy for one woman in particular, he told me, "an older lady who clearly did not want to go out there and undress. She kept coming back into the building. She really felt troubled, but eventually she went." A front-page photograph in the *Kansas City Star* the next day underscored the sense of distress caused by the ordeal. A woman with an anguished expression, wrapped only in a blanket, is shown being escorted to the bus. She is walking barefoot in the icy outdoors with hair soaked from the decontamination washing.

A few days after the event, Christine Vendel, a *Kansas City Star* reporter who covered the incident, received a fax from one of the firemen. He expressed anger about having been forced to undergo the decontamination washdown in the nude and by a woman. Vendel told me that he urged her to write about his being a victim of sexual harassment, but she demurred. None of the three members of the fire department whom I interviewed acknowledged knowing that anyone had sent such a fax.

The event prompted potentially serious health problems for two women who had to be taken to the hospital. After decontamination, one complained of breathing difficulty and the other became unconscious from an allergic reaction to the bleach solution. The psychological impact on many victims was profound. Toni Blackwood, chair of the local Planned Parenthood Board, lamented that patients and staff had been "traumatized . . . and terrorized."

Rebecca Poedy, financial officer at the Planned Parenthood clinic in Boise, Idaho, opened a letter on Friday afternoon, February 26, 1999. She immediately took it to the clinic director, who called 911. Police detectives were the first to arrive and told her to wash her hands and face. Soon after, with the arrival of medical and fire emergency responders, "there were people all over and it was kind of jumbled," Poedy told me. "No one seemed in charge. One of them said to send the staff outside, but others said no, stay in the building."

Initially, the 24 people in the clinic were kept inside but were then directed outside to the parking lot. After an hour they were brought inside again and were quarantined for another 3 hours. Meanwhile the street around the clinic was blocked off. For more than 3 hours, people in other offices in the area complex were not permitted to leave. Poedy estimates that around 150 were detained.

Poedy had been kept separate from the others and said that when firefighters in masks and full gear came for her, "I really began to worry." They had put up a makeshift shower and wading pool in the middle of the street. A tarp was strung up above the shower, reaching down to her knee level. She was instructed to go under the tarp, disrobe, slide her clothes out, and shower. After washing with soap and "very cold water," she had to don a white suit, gloves, and surgical mask. An ambulance had been summoned and paramedics insisted on placing her in a bodybag before taking her to the hospital. "I objected to the bodybag, but they said I had no other option. They zipped it up to my neck." At the hospital a doctor told her that if she had contracted anthrax, there might be no cure.

Still in mask and gloves, Poedy was made to lie in the bodybag for 3 hours. She was allowed out only after word was received that initial tests found no evidence of anthrax in the letter. When the shaken Poedy returned to the clinic, Mary McColl, the office's chief executive officer, became distressed about how upset Poedy appeared. McColl had been among the 24 people in the building who had been kept isolated. Months after the incident McColl still expressed her chagrin about Poedy's treatment and her impression that no one seemed to be in charge.

On Monday, April 19, 1999, a sheriff's deputy found a note in the parking lot of the county courthouse in Biloxi, Mississippi. It claimed that anthrax had been placed in the building's ventilation system. About 100 people were evacuated, and the building remained closed until the next day.

The public read about the incident 2 days later in the local newspaper, the *Sun Herald*. Health and law enforcement officials were cited as doubting that anyone had been exposed to anthrax. But an unnerving warning appeared in the story: "As a precaution, however, health officials are warning people who were at the courthouse Monday to be on the lookout for flulike symptoms such as a cough, runny nose or fever, and to contact their doctor if they have concerns." According to Blake Kaplan, the reporter who wrote the story, no one knew how many people had been in the area or how many ended up concerned if they had a cough or runny nose. He did say that not until a week later was final confirmation received that no harmful biological agents had been present.

A month later, on Monday, May 17, Robert Duffy was opening mail in the alumni office at Le Moyne College in Syracuse, New York. When Duffy, assistant director of the office, unsealed a slightly bulging envelope, a white powder puffed out onto his clothes. "The letter was all folded up so I had to paw around in the envelope and got the stuff all over my hands," he said.

The letter said, "Anthrax. Congratulations. Death to the Catholics who oppose abortion," Duffy told me. Duffy instructed his assistant to move away and call security; then he locked himself in his office. After the police arrived, he was told to remain in his office while they evacuated approximately 30 other people to an area outside. A Catholic high school across town had just received a similar threat letter and HAZMAT crews were busy over there. Duffy and the people who were detained outside had to remain in place for 3 hours until the high school was secured. During that time Duffy was on the telephone answering calls from the FBI, county health officials, and HAZMAT experts. He described the powder to them as "coarse and grainy, like ground-up chalk." Their answer was, "Yes, that's what anthrax looks like."

Panicky, Duffy went on the Internet to learn more about anthrax, but the information made him more anxious. "When I read about 90 percent fatality rates, I stopped reading." Three hours later paramedics arrived in protective gear and masks and told him to undress and wash. Duffy was then dressed in a mask and protective outerwear and escorted to an ambulance. His assistant was already in the ambulance, and the two were brought to a hospital where they were scrubbed with soap and water and told to wait. After a few more hours they were informed that tests indicated anthrax almost certainly was not present.

In describing the event to me a month after it happened, Duffy rued the delayed response. "I was in the office for 3 hours, waiting," he said. "I think the emergency people should have responded to me sooner." He also regretted that authorities did not later contact victims to help with psychological problems brought on by the ordeal.

The stunning increase in the number of anthrax hoaxes was underscored at a congressional hearing in May 1999. A statement by Robert Burnham, chief of the domestic terrorism section of the FBI, indicated that in 1997 the bureau had opened 22 cases involving the "threatened use or procurement of . . . biological materials with intent to harm." In 1998 the figure was 112. Between January and May 1999 the number had already reached 118. Neil Gallagher, assistant director of the FBI's national security division had observed: "Not a day goes by without us hearing from somewhere in the United States about an anthrax threat."

The spate of hoaxes began after several months of publicity about the horror of germ weapons, anthrax in particular. Media coverage of the B'nai B'rith incident in April 1997 had been extensive. But the most widely reported event that year concerning anthrax was the display of a bag of sugar by Secretary of Defense William Cohen on national television. A November 17 *New York Times* column said that he "scared everyone by going on ABC's 'This Week' and holding up a 5-pound Domino sugar bag to show the small amount of anthrax . . . it would take to send half of Washington into writhing death throes." The secretary's performance was repeatedly shown on other programs and broadly reported in the press. It received at least three repetitions in the *Times* itself. On November 23, November 30, and December 16, articles recounted Cohen's vivid show-and-tell. Thus, for more than a month the most influential newspaper in the United States was regularly reminding readers of Cohen's frightening anthrax metaphor. (Two years later the secretary's sugar display and his "half-of-Washington" threat could still be seen both on ABC and government Web sites.)

Then in February 1998 two men were arrested in Las Vegas, on suspicion of possessing vials of anthrax. One of them, Larry Wayne Harris, had been found guilty of mail fraud in 1995 after ordering bubonic plague bacteria from a microbiology supply company. They were rumored to be traveling east to release the germs in the New York City subway. The story later proved false, and the vials turned out to contain a harmless veterinary anthrax vaccine, but articles about bioterrorism and the public's vulnerability were rampant. Later, other officials, including President Clinton, were issuing their own alarming pronouncements. While saying he did not want to cause "unnecessary panic," Clinton told a reporter that more than any other threat the possibility of a germ attack "keeps me awake at night."

The hyperbolic statements were intended to rouse people, especially other decision makers, from complacency about the threat of biological weapons. But the statements created another risk. They heightened the chances that unsavory people would get ideas they would not otherwise have had. The drumbeat of publicity about anthrax hoaxes, coupled with official warnings, appears to have had that unpleasant effect. The dramatic increase in the number of hoax incidents seemed to parallel the increased media attention to the subject.

We are left with a difficult paradox. Until the 1997 B'nai B'rith anthrax hoax, bioterrorism threats by mail had been almost unheard of. But subsequent copycats threats, and the disruptions they prompted, were widely reported. Along with well-intended warnings by government leaders, they generated an epidemic of false threats. Could the idea for the real anthrax letters in 2001 have been seeded by the spate of hoax letters and the publicity about them? The possibility poses an agonizing dilemma. It challenges the treasured ideal of unfettered reporting in the face of life-threatening consequences.

Just who was making those hoax threats?

Bloomington is the home of Indiana University, the state's flagship institution of higher education. The town also gained notoriety on November 10, 1998, when Bloomington High School South's 1,800 students were evacuated from the school building. The order came after the principal, Jim Rose, opened a note containing powder and a message that he had been exposed to anthrax. The threat turned out to be a prank by two 16-year-old boys. Convicted in juvenile court of conspiring to make a false report, the boys were put on probation and ordered to perform 80 hours of public restitution work. (In the days before the prank, highly publicized anthrax threats had been received at a Catholic church and an abortion clinic in Indianapolis.)

Five weeks later, on Friday morning, December 18, 1998, an anonymous telephone call to the Bankruptcy Court in Woodland Hills, California, warned that anthrax had been released in the building's air-conditioning system. About 150 firefighters, police, medical responders, and federal agents arrived on the scene and evacuated the building. Some 90 courthouse employees were quarantined in the parking lot for 8 hours until the building was deemed safe.

One of the canceled court hearings that day involved charges that Harvey Craig Spelkin had failed to attend earlier proceedings brought by a former employer. Spelkin, 53, was an accountant from Calabasas, outside Los Angeles. The employer alleged that Spelkin had embezzled $100,000 from him. In the course of an interview by the FBI, a court staffer mentioned Spelkin's past absences and, when confronted, Spelkin admitted making the telephone call.

Following his conviction in early 1999 for threatening to use a weapon of mass destruction, Spelkin was sentenced in July to one night in jail and 400 hours of community service. He was also ordered to pay more than $600,000 to cover the costs of the police and fire department operations. Spelkin's effort had taken place amid a rash of anthrax hoaxes in the Los Angeles area that had received heavy newspaper and television coverage.

Newark, California is on the east side of San Francisco Bay, across from the better-known Palo Alto. About 1:30 p.m. on December 29, eleven days after Spelkin's publicized phone call, a 911 call was made to the Newark police station. "There are 400 people in this building and anthrax has been released," the anonymous caller said. The call was traced to WorldPac, a Newark auto parts shipping company. HAZMAT crews responded immediately but could find no evidence of anthrax. When police and FBI agents played a tape of the conversation to the company's supervisors, they recognized the voice as belonging to an employee, Robert Alinea Peterson.

During an interview with the FBI, Peterson admitted to making the call. His reason? He wanted to leave work early but was concerned about losing his job because of his past absenteeism. Months later, in a plea bargain, Peterson was placed on probation. According to court documents, he had been inspired to make the anthrax threat by the extensive coverage of recent anthrax hoaxes, especially in California. In a February 8, 2001, article in the Toronto *Globe and Mail*, Gary Ackerman and Jeffrey Allan noted that anthrax hoaxes in the United States seemed to be coming in waves. "This suggests," they wrote, "that the media's reporting of an initial event may lead to copycat hoaxes."

While these hoaxers made one-time threats, others have been responsible for multiple scares. On January 12, 2000, the *Milwaukee Journal Sentinel* reported on a spate of incidents:

> Two more anthrax threats were sent to [Planned Parenthood] offices in Milwaukee and Racine on Wednesday as such hoaxes continued to sweep across the country, authorities said. The circumstances were similar to incidents earlier this week at three other Milwaukee sites, as well as at a middle school and children's services agency in Kenosha. In each instance, letters contained a powder with a note threatening that the substance was anthrax.

The targeted establishments were as varied as those in the rest of the nation. One was Bullen Middle School in Kenosha, where 800 students were evacuated. Another was the Community Adoption Agency in Manitowoc, where a letter warned: "This is anthrax—you will die." Yet another was to the Affiliated Pregnancy Counseling Service in Racine, which was described in the *Journal Sentinel* as "an all-volunteer, non-profit Christian agency that opposes abortion."

FBI special agent Brian Manganello, who was based in Milwaukee, was frustrated. "It's running the gamut just like last year, abortion clinics, schools, hospitals, health care providers, pro-life agencies. It doesn't seem to be a sensible pattern," he said. Then came a break in the Milwaukee-area cases. Toward the end of the month, police announced they had arrested Mickey Sauer, 25, a resident of Kenosha. He was charged with mailing 17 anthrax threats to area targets. He later pled guilty to sending the letters between January 5 and 18, 2000. In September he was sentenced to 21 months in prison. His motive was not reported. A Kenosha police officer described him to me simply as "a local nut."

Anthrax hoax threats continued apace into the fall of 2001. Despite periodic announcements that a perpetrator had been apprehended, most cases remain unsolved. Out of some 40 indictments there had been only 20 convictions. Moreover, the nature of the targets continued to vary. Although abortion clinics were disproportionately targeted, they were a minority in the overall pool that included antiabortion groups, nightclubs, churches, schools, stores, hospitals, post offices, courthouses, news media offices, even FBI offices.

The precise number of anthrax hoaxes between the time of the B'nai B'rith incident in 1997 and the anthrax letters in the fall of 2001 was unclear. After mid-1999, when an FBI spokesperson reported a rate of a couple hundred a year, the bureau did not release specific figures. Subsequently, media coverage of hoaxes also declined, in part because they were no longer "news." But after October 2001 the situation changed. With the discovery of actual spores in the mail, bioterrorism threats of all sorts were receiving heightened attention. Newspapers again were giving widespread coverage to anthrax hoaxes. Indeed, there were many more of them to cover. On November 8, 2001, Senator Joe Biden announced that in the previous 8 weeks the FBI had responded to more than 7,000 suspicious anthrax letters.

Jason Pate, an analyst with the Monterey Institute of International Studies, believes the actual number of anthrax hoaxes has been far fewer than the FBI's figures. He heads a project that monitors reports of terrorism incidents and says that in the wake of the anthrax letters there may have been tens of thousands of false alarms. But among them were "perhaps only 1,000 actual anthrax hoaxes/threats worldwide." He explained the discrepancy:

> The FBI counts any investigation or call or report as a "hoax," where we [at Monterey] only count a hoax as an event where there was a letter/powder/threat that indicated that someone was trying to create an incident. What we call "false alarms," including the thousands of calls that someone had seen "white powder" or a "mysterious cloud," law enforcement agencies and the FBI often label hoaxes.

Whether the number is many thousands, as the FBI suggests, or about 1,000, as Pate believes, it is clear that the number of hoaxes mushroomed after the anthrax letters in 2001.

Generalizations about the hoax perpetrators according to motive or personality are elusive. But the characteristics of at least one hoaxer seem oddly similar to those of other notorious terrorists.

Among the hoax culprits who have been found, Clayton Lee Waagner is distinctive and not only for the huge quantity of threat letters he sent. Unlike the mix of pranksters and ne'er-do-wells who sought aggrandizement for themselves, he terrorized in the name of a larger purpose. Waagner holds a world view that justifies killing people for a deeply held cause—to save the "unborn." His motivation resembles others who murdered for an ostensibly higher purpose. Theodore Kaczynski, the Unabomber, killed with mailed bombs as a statement against modern technology. Timothy McVeigh, the Oklahoma City bomber, destroyed a federal building and its occupants because he despised the U.S. government. Mohammed Atta led the airline attacks on the World Trade Center and the Pentagon in opposition to democratic and Western values.

Soon after September 11, 2001, Dr. Jerrold Post interviewed Atta's professors at the University of Hamburg. Post is an expert on political psychology at George Washington University. Atta's teachers told Post that until Atta visited Osama bin Laden in Afghanistan, he had been "dedicated to his studies, was polite, engaged, bright." Like Kaczynski and McVeigh, a middle-class intelligent young man became a murderous zealot.

Clayton Waagner proudly labeled himself a terrorist on the Web site of the Army of God in June 2001, five months before he sent the letters. Moreover, he threatened death to people who failed to respond to his admonition to cease abortion activities. Dennis Roddy, a reporter for the *Pittsburgh Post-Gazette*, has interviewed and corresponded with Waagner. In an e-mail to me, Roddy spoke of Waagner with awe: "Clayton is very smart, very clever, self-educated and if I were going to steal a car, I would want him along for the advice. His capacity for staying hidden and keeping himself afloat on the run is utterly amazing."

Roddy is not alone in his reaction. A government attorney who has worked with Waagner found him no less compelling. The attorney had spoken about him with his previous lawyer and with judges who presided over his case. "They all found him impressive," said the attorney, who requested anonymity. The attorney described Waagner to me as "an extremely nice man in terms of his conduct, extremely polite. He is a gentleman to work with." As the attorney warmed to his subject, he became even more effusive: "Waagner is very clear about what he has done. In that sense it's refreshing. He truly believes in his cause. Look, I consider myself very liberal and I don't believe what he believes in. But after you talk to him for a while and hear his sincerity, you almost question yourself."

Was this attorney overlooking the terror that Waagner had inflicted? Does he believe that Waagner's charm and sincerity can mitigate his illegal conduct? Before I raised these questions, the attorney hauled himself back to hard reality. He chuckled, "Then when you're away from him and you think about what he's done, you realize, 'No, no.' He's very clever, but very misguided."

In March 2003, responding to my mailed request, Clayton Lee Waagner, 46, phoned me from the Lewisburg Penitentiary. "Please call me 'Clay,'" he began, in an airy, friendly voice. His mature use of language was all the more notable for his having dropped out of school before reaching the 10th grade. After altercations with his family, at age 16 he moved out of his Georgia home and began to live on his own. The next year he joined the Coast Guard for a short stint— "I did well in their test to get in." He worked for the Christian Broadcasting Network as an engineer and chauffeur but left because "I got tired of it." He became a computer programmer but quit because "I didn't like it." He moved to Alaska and ran a commercial boat, "but that didn't work out."

Waagner, who has been married since he was 20, eventually made his way with his wife and eight of their nine children to western Pennsylvania. In 1999 they moved into a house in Kennerdell, a rural community about 50 miles north of Pittsburgh. Around that time, his oldest daughter, Emily, who was married and the mother of a child, suffered a miscarriage. Until then he said, "I was against abortion but thought it was wrong for me to tell others what to do." But seeing the 5-month-old fetus ignited something in him to "take up weapons against abortionists."

Before we spoke by phone, I had received answers from him to several questions I posed in a letter. Waagner's natural intelligence comes through in his handwritten responses, dated February 27, 2003. Excerpts:

Q: Where did you get the idea of making anthrax threats by mail? Was it from publicity about the actual anthrax letters sent in the fall of 2001 or from publicity about anthrax hoax letters before then?

Waagner: The idea came to me the minute it was announced that the fellow with American Media died from mailed anthrax. At that time I had not heard of fake anthrax threat letters. Ironically, I thought using the fake powder against abortion clinics was original. Eighty such letters had been sent to clinics in previous years.

Q: What exactly was the powder you were using as a mock anthrax agent? Where did you get the powder?

Waagner: On the first mailing (10/12/01, 498 sent—280 received) I used white flour purchased from an Atlanta area Kroger. Those were sent via U.S. mail. Those closed the clinics for a week as in October, few local areas had anthrax field test [kits].

The second Federal Express mailing in November (11/8/01, 298 sent—298 received) I used bacillus thuringiensis (BT), which is a Monsanto product used to kill bole [sic] worms off cotton. BT is harmless to humans, but identical in every other way to anthrax. At this time most areas had anthrax field test [kits] in place. When the BT caused a positive response for anthrax the clinics were again closed for a week. It took CDC a week to tell the difference. BT even grows a culture like anthrax, but after 5 or 6 days the difference is manifest.

BT in powder form is hard to find. Generally it is applied by aircraft so it's sold in liquid form. Finally I found it powdered at a small feed store in rural South Carolina. I bought 2 four pound bags for $2.66 each. When I asked to make sure the stuff was harmless, the old fellow replied, "You could make biscuits out of the stuff and eat it."

Q: When you were apprehended at the Kinko's, were you in the store? What services were you using in the store?

Waagner: I had been inside the Kinko's using the computer. I was looking up the fax numbers of abortion clinics on the Internet for my next attack. When I saw two police officers . . . walk in the store I left. One of them followed me out and asked me for id [identification]. She was 5'2", 110 pounds. They had no idea who I was (FBI 10 Most Wanted) and I was arrested without a gun being drawn.

Waagner's answers are impressively phrased, if not always factually accurate. But the nonchalance with which he describes his activities and his plans for another attack is chilling. Moreover, his reference to making sure "the stuff was harmless" is at odds with the intent he expressed elsewhere. During his taped conversation with Neal Horsley, just before he was apprehended in December 2001, Waagner said: "This anthrax scare that I did to the abortion clinics . . . I didn't do the real anthrax. I don't have real anthrax. If I did it would have been in those letters."

Horsley responded: "Right, you would've killed—."

"Oh, without a doubt."

Of what relevance are Waagner's motives and actions to the real anthrax letters? Waagner is not even remotely suspected of having a connection to the killer letters. But as he readily proclaims, he is a terrorist nonetheless. Perhaps the most unnerving part of his performance was the demonstrated ease with which he sent at least 550 letters (778 by his count) laced with bogus anthrax. Had the letters contained *Bacillus anthracis* rather than *Bacillus thuringiensis*, the consequences could have been catastrophic. Instead of the 22 cases of anthrax, including five deaths that resulted from the poison letters, extrapolations would raise the toll to 2,200 victims and 500 deaths. In fact, Waagner's persona and motivation seem to have less in common with most fellow hoaxers than with killer terrorists.

CHAPTER
NINE

WHO DID IT?

M is restless with anticipation. It is Monday, September 17, and the day has been carefully planned. Late in the afternoon, in his small makeshift laboratory, he takes an 8-ounce jar containing a light tan powder from the shelf above the workbench. Composed of microscopic particles, the powder resembles exceedingly fine flour. Even though he has been vaccinated against anthrax, M is very cautious. Wearing gloves, jumpsuit, and face mask, he gingerly unscrews the cap.

He taps some material into a small plastic receptacle perched on an electrically calibrated scale. The scale is hooded to avoid spilling. When the scale registers 1 gram, he stops. He sifts the weighed powder into a transparent plastic test tube. The powder is so light that he must hold the tube still for several minutes as the material slowly settles into it. He repeats the exercise with a second test tube.

Each tube now contains billions of particles. The particles are anthrax spores, some clinging to each other in clusters of 20 or 30 because of electrostatic attraction. But the spores have been treated with a silica-based material that minimizes the electrostatic effect, and many particles are individual free-floating spores. A single spore, about 1 to 3 microns in length, is so tiny that 10,000 could fit on the head of a pin.

Days earlier M knew he would send identical terror messages to the editor of the New York Post and to Tom Brokaw at NBC in New York City. In preparation, he purchased 34-cent prestamped enve-

lopes produced by the U.S. Postal Service and printed their addresses on them in block letters. He also printed a message on a single piece of paper and made copies of it on a copy-machine:

09-11-01
THIS IS NEXT
TAKE PENACILIN NOW
DEATH TO AMERICA
DEATH TO ISRAEL
ALLAH IS GREAT

As evening approaches he folds a copy of the message into each envelope. Then he slowly empties the powder from the first test tube into one envelope and from the second tube into the other envelope. Careful not to spill any material on the outer surfaces, he seals the envelopes with tape. He places them in a plastic Ziploc bag, then in a brown paper bag, and rests the bag on the workbench. He sits silently through the evening.

By midnight the streets in downtown Princeton are quiet. The night sky is clear, the temperature a comfortable 50°. As rehearsed during previous evenings, M drives at deliberate speed south on Nassau Street. The Princeton University campus is on the left as he crosses Witherspoon Street and then John Square. Three short blocks later he makes a right on to Bank Street, pulls over and parks. Before leaving the car, M draws on a pair of latex gloves and reaches for the Ziploc bag next to him. He glances about, the bag firmly in hand, and he walks several yards back to Nassau Street where three blue mailboxes stand near the curb. They are opposite the American Express travel office on the corner.

No one else is near. Good. September 11 comes to mind and M thinks: "That day was the first step. Anthrax will be the second." Standing in front of the boxes, he unzips the plastic bag and pulls out the two powder-filled letters. He deposits them in the middle box, quickly returns to the car, and drives off.

At 11 a.m. the next morning a postal carrier removes a white plastic tub of mail from each mailbox and replaces each one with an empty tub. After picking up mail from other boxes along his route, the carrier brings his collection to the local post office in Princeton. In the afternoon the accumulated batches are transported to the sorting and distribution center in nearby Hamilton. Meanwhile, spores have begun to leak from the anthrax letters onto other mail. The cross-contamination is intensified at the Hamilton facility by the turbu-

lence of the sorting machines. As the letters are routed to their destinations, they continue to leak spores along the way. Some time later they are delivered to the offices of the addressed parties. The anthrax attack is under way.

I devised this vignette, purposely ignoring the presumed motives of the perpetrator, and asked some experts if they thought it plausible. The presumed presence of silica material with the spores was questioned by some. It is an issue of continuing disagreement among scientists. Further, Barbara Hatch Rosenberg said, "I do find an implausibility." Responding by e-mail, she wrote: "Insufficient containment (unless you are making the case that the first mailing was not easily aerosolizable, unlike the second). The Daschle anthrax jumped off the microscope slide when they tried to look at it." "Otherwise," she said, "OK."

Her observation about the aerosol characteristic is important. The quality of anthrax in the letters sent to NBC and the *New York Post* apparently was different from that in the letters sent to Senators Daschle and Leahy. The first two letters, which were postmarked September 18, 2001, reportedly contained dead vegetative anthrax organisms and other debris mixed with spores. The second two, postmarked October 9, contained a pure preparation of spores that more readily became aerosolized. That is, they easily floated in the air— "jumped off the microscope slide," as Rosenberg said. (In a letter to the *New York Times* on May 18, 2002, Federal Bureau of Investigation [FBI] spokesman John Collingwood confirmed the difference in potency between the two pairs of letters.) Thus, my scenario would seem appropriate for the September mailing, which is the one I depict, but not for the October mailing.

When I asked Rosenberg to elaborate on what she thought sufficient containment would entail, she responded that the perpetrator either "had access to a containment lab or he improvised one on his own." He had to be "completely protected," she said. The material somehow had to be isolated "or else he had to be working where contamination would never be found."

Rosenberg, a professor of environmental science at the State University of New York in Purchase, has been working on biological weapons issues since the early 1980s. For her, as for many in both the law enforcement and scientific communities, available information suggested a lone perpetrator who had worked in a U.S. government laboratory.

In January 2002, Tom Ridge, director of homeland security, acknowledged that immediately after the initial anthrax incidents "our natural inclination was to look to external terrorists." But now, he said, the "primary direction of the investigation is turned inward." In fact, as early as October 25, 2001, George Tenet, director of central intelligence, and Robert Mueller, FBI director, told senators they were not ruling out any possibilities, but that an Iraqi connection seemed unlikely.

Days later news reports suggested the focus had become domestic. The area around Trenton and Princeton, from which the anthrax letters were mailed, appeared to be of particular interest. On October 31, the *Wall Street Journal* reported that investigators were interviewing laboratory researchers in the Princeton area, asking about "specific equipment and whether any such specialized machinery has disappeared." Then, on November 9, the FBI posted a profile on the Internet that strongly suggested a domestic perpetrator. The message was unusually specific about the characteristics of the presumed culprit. Tracey Silberling, a spokeswoman for the bureau, acknowledged that the FBI had never before made public such extensive material on an unsolved case. She said that the information would educate people about the threat and perhaps "ring a bell with someone" who might then contact the bureau.

Under the heading "Amerithrax Press Briefing" (Amerithrax was the FBI's name for the investigation), the bureau offered "linguistic and behavioral assessments" of the anthrax mailer. The notice asked the public "to study these assessments and reflect on whether someone of their acquaintance might fit the profile." The anthrax mailer, according to the notice, was likely an adult male with a scientific background "or at least a strong interest in science." He has access to a source of anthrax and knows how to refine it, and he is familiar with the Trenton, New Jersey, area. (Princeton is 10 miles north of Trenton.) Further, the perpetrator is "a non-confrontational person, at least in his public life." But he "may hold grudges for a long time vowing that he will get even with 'them' one day." And he "prefers being by himself more often than not."

The personality characteristics resembled those of Ted Kaczynski, the Unabomber. Between 1978 and the time of his capture in 1996, Kaczynski was responsible for 16 bomb attacks, most of them by mail, that killed three people. The experience was still fresh in people's minds. James Fitzgerald, an FBI profiler, told the *Los Angeles Times* that the Unabomber's profile helped inspire the bureau's description of the presumed anthrax mailer.

Three days after the FBI notice, on November 12, Barbara Hatch Rosenberg posted her own assessment of the anthrax perpetrator. It appeared on the Web site of the Federation of American Scientists (FAS), an organization of 3,000 scientists whose focus is on science policy. Rosenberg chairs the federation's working group on biological weapons. Her message, and others that she posted in subsequent months, catapulted her into public visibility. In April 2002 an admiring article by Lois Ember in *Chemical and Engineering News* characterized her as an "intellectual provocateur." At the same time, for David Tell, opinion editor of the *Weekly Standard*, she was "Miss Marple" with a "crackpot theory."

Rosenberg's message moved the FBI's presumptions into new territory. She asserted that the perpetrator was an American scientist with access to U.S. "weaponized" anthrax or had been taught by an expert how to make it. And she speculated that if the bureau believed "the anthrax was an 'inside' job, it would probably want to cover it up." Failure by the FBI to identify the perpetrator, she wrote, might have arisen from a desire to protect politically embarrassing information.

Barbara Rosenberg is a demure enthusiast. Fashionably dressed, behind dark-rimmed glasses, her appearance hardly suggests that of an impassioned crusader. But she is intensely devoted to keeping the world free of biological weapons. She received a Ph.D. in biochemistry from Cornell University Medical School in 1962. Not until the 1980s, while working at the Memorial Sloan-Kettering Cancer Center in New York, did she turn to biological warfare issues. During that time, she authored a blistering environmental critique of a proposed biological research laboratory at the Army's Dugway Proving Ground in Utah. She also began to lead FAS efforts to enhance the effectiveness of the Biological Weapons Convention.

Now, with her Internet posting in November, she turned her attention to the anthrax letters. What was the basis for her assessments? I asked her. "I had inside sources," she answered. "I heard from people in the biodefense community who thought a specific person was the most likely one who did it." There was no proof at the time, just strong suspicions. She emphasized that her analysis was not based on the FBI's profile but on her own sources and deductions.

Rosenberg refined her assessments and posted them on December 10, 2001. Now she suggested that the government had "undoubtedly known for some time that the anthrax terrorism was an inside job." A month later, on January 17, another posting: "By now the FBI must have a good idea of who the perpetrator is." And she introduced an ideological component: "The choice of Senators Daschle and Leahy [as targets] suggests that the perpetrator may lean to the political right and may have some specific grudges against those Senators." This conjecture does not, of course, explain why anthrax letters were sent to the American Media tabloids or to the right-leaning *New York Post*.

Rosenberg's January notice included a section titled "Possible Portrait of the Anthrax Perpetrator." The perpetrator, she wrote, has a "doctoral degree in a relevant branch of biology"; he "works for a CIA contractor in [the] Washington, DC area"; he "knows Bill Patrick [the former U.S. chief bioweaponeer] and has probably learned a thing or two about weaponizing from him"; he "has had a dispute with a government agency." While not embracing Rosenberg's specificity, the FBI also remained fixed on the domestic loner theory. At the end of January, Van Harp, assistant director of the bureau's Washington field office, sent an extraordinary letter to the 32,000 U.S. members of the nation's leading organization of microbiologists:

> I would like to appeal to the talented men and women of the American Society for Microbiology to assist the FBI in identifying the person who mailed these letters. It is very likely that one or more of you know this individual. . . . Based on his or her selection of the Ames strain of *Bacillus anthracis* one would expect that this individual has or had legitimate access to select biological agents at some time.

The letter concluded with a reminder that the reward for information leading to the culprit's arrest and conviction had grown to $2.5 million.

Rosenberg viewed the FBI's mailing as a waste of time. All but a couple of hundred members of the society, she said, live in a "different world" from the perpetrator. Although she would not name the suspect, she was impatient with the FBI for not doing so. The evidence was sufficient to "single out the perpetrator from the other likely suspects," she wrote. But the bureau's reluctance to name him was perhaps because he knew something "sufficiently damaging to the United States to make him untouchable by the FBI."

Rosenberg's assessments were widely reported, often in a political cast. Liberal magazines, including the *American Prospect* and the *New Yorker*, published articles that sympathized with her position. Conservative publications, notably the *Wall Street Journal* and the *New York Post,* were skeptical. No journalist was more fixed on the issue than *New York Times* columnist Nicholas Kristof. Between January and August 2002 he devoted six columns to the subject. Incorporating most of Rosenberg's themes and then some, he criticized the FBI for "lackadaisical ineptitude in pursuing the anthrax killer."

David Tell, in the *Weekly Standard*, included a point-by-point critique of the argument that the anthrax mailer was a domestic scientist. Tell was as critical of the FBI as he was of Rosenberg. The result was an ironic triangle: He and Rosenberg pointedly disagreed with each other, and both disagreed with the FBI, albeit for different reasons.

Whereas Rosenberg contended that the Ames strain in the letters likely came from an American laboratory, Tell said that the United States had shared the Ames samples with labs overseas, which could have been the source. Whereas the FBI believed that the block capitals in the anthrax letters were written by a native English-speaker, Tell asked why the writer might not have been "someone who grew up speaking a language—like Arabic—whose alphabet has no upper or lower cases." Whereas Rosenberg thought the writer's warning to take penicillin meant he did not intend to kill, Tell disagreed. He believed that no "benignly inspired" expert would now prescribe penicillin for an anthrax infection. (Cipro and doxycycline are considered the antibiotics of choice.) He concluded that "the possibility is far from foreclosed that the anthrax bioterrorist was just who he said he was: a Muslim, impliedly from overseas, who thought the events of 09-11-01 were something to be celebrated—and who would have been doubly pleased to see 'you die now.'"

In 2003, Rosenberg remained as convinced of her case as she was in 2001. "To my mind, everything that subsequently became known played right into my hypothesis, which strengthened my own convictions," she said to me. And she is equally steadfast about the unflattering criticism she has received: "It rolls right off. I mean, the people who are making these statements are right wingers who have axes to grind. They're trying to prove that it was Iraq, or Al

Qaeda, or anything but the U.S. biodefense program. So, I mean, I expect that." When I noted that the person she described in her postings seemed to fit the profile of Steven Hatfill, she responded softly, "I have not mentioned a name and I don't intend to."

Pat Clawson exudes gusto and strong opinions. As he climbed out of his dented 1988 red Plymouth Reliant, he announced, "I'm just a fat, ugly Irishman who doesn't like what's happening to my friend." It was the end of January 2003, a cold but clear afternoon in Washington. We found a quiet table at Kelly's Irish Times, a downtown pub. Over a corned beef sandwich and some Ellis Island beer, for 2 hours Clawson catalogued the injustices he believes Steven Hatfill has suffered. During the 1990s he and Hatfill developed a friendship at dinner parties attended by a circle of fellow conservatives. Both men are the same age, 48, and they share a world view that they purvey with spirited confidence.

When I spoke with Clawson, Hatfill had been unemployed for 5 months. Clawson, a veteran radio and television reporter at NBC and CNN, had also recently left his job with Radio America. Unlike his friend, he anticipated finding work again quickly. Hatfill has been instructed by his lawyers not to talk to any outsiders, but he has accepted the media-savvy Clawson's offer to act as his spokesman. With Hatfill's blessing, Clawson has regularly appeared on the air and in print on his behalf.

Clawson first learned that Hatfill might be in trouble on June 25, 2001. He heard a radio reporter say that the FBI had searched an apartment in Frederick, Maryland. "It was connected with the anthrax investigation and he mentioned 'Steven Hatfill,'" Clawson said. For a moment Clawson wondered if Hatfill could have been the anthrax killer. But then the broadcaster indicated that Hatfill was not a suspect, and had consented to the search. "Well, hell," Clawson thought, "Steve's just being a good soldier and cooperating with the FBI."

Clawson knew that Hatfill was an expert on bioterrorism and recalled Hatfill telling him the FBI had interviewed him about the anthrax letters. No cause for concern. He was among hundreds being interviewed, including almost everyone who had recently worked at the U.S. Army Medical Research Institute of Infectious Diseases (USAMRIID). But then, according to Clawson, media stories began suggesting that Hatfill had become the center of the investigation. Clawson cited the articles by Nicholas Kristof and others. "Steve was being portrayed as a nutty, lone scientist who was pissed off at the government because he lost his security clearance, lost his job, and was a closet racist who had worked for racist regimes in Rhodesia and South Africa." Clawson's baritone inflections sound remarkably similar to those of talk-radio's Rush Limbaugh. Clawson paused, rested his beer on the table, and said, "That wasn't the Steve that I knew."

Clawson reached Hatfill and asked him what all these stories were about. "Pat, I don't know what the hell is going on," Hatfill answered. Clawson watched his friend break into tears. "They're following me around the clock, everywhere I go, and I don't have a damn thing to do with any of this."

Hatfill's background is certainly unusual. He had distorted items on his résumé, including a false claim that he held a Ph.D. But much about his record is not in dispute. He was born in St. Louis in 1953, grew up in Illinois, and graduated from Southwestern College in Kansas in 1975. After serving in the U.S. Army, he went to Rhodesia (now Zimbabwe) to study medicine. In 1984 he received an M.D. from the Godfrey Huggins School of Medicine (now the University of Zimbabwe). He completed a hematology residency in South Africa, where he also obtained a master's degree in medical biochemistry and another one in microbial genetics.

In 1994 he submitted a Ph.D. thesis on molecular biology at Rhodes University in South Africa but never received the degree. Hatfill left Africa in 1995 and spent a year as a research scientist at Oxford. He returned to the United States for a 2-year fellowship at the National Institutes of Health and then 2 years, from 1997 to 1999, at USAMRIID. There he conducted research on viruses, including Ebola and Marburg, which are considered possible biowarfare agents. Afterwards, in January 1999, he began to work for Scientific Applications International Corporation (SAIC), a private defense contractor in Virginia.

Hatfill was fired from SAIC in March 2002, though Clawson refused to say why. According to media reports, it was because the Department of Defense had suspended his security clearance in August 2001 and Hatfill's efforts to regain the clearance had been unsuccessful. The suspension might have been related to his participation in the 1980s with the military in Rhodesia and South Africa. An intelligence analyst who knows Hatfill confirmed to me that Hatfill had been involved with special operations there of an unspecified nature. In any case, after leaving SAIC, Hatfill was hired to work in a biomedical training program at Louisiana State University. But amid the notoriety, he was let go in early September 2002.

Hatfill's intense interest in biodefense is obvious from his résumé. In the late 1990s he developed a biological warfare syllabus for emergency room physicians and he was a biological weapons consultant for the Washington, D.C., Metropolitan Medical Strike Force. His résumé also says: "Working knowledge of the former U.S. and foreign BW programs, wet and dry BW agents, large-scale production of bacterial, rickettsial, and viral BW pathogens, stabilizers and other additives."

These were among the reasons Hatfill had been targeted, implicitly by Rosenberg, and explicitly by Kristof and Attorney General John Ashcroft. The search of Hatfill's apartment in June was followed by another search on August 1, but this time the FBI brought a criminal search warrant. While still denying that Hatfill was a suspect, the FBI and the attorney general deemed him a "person of interest." He was one of 30 persons of interest, according to the bureau, but Hatfill apparently was the only one under sustained FBI surveillance.

Clawson was convinced that Hatfill's public silence was not helping him. "Steve, you've got to get your side of the story out. You need to talk to the press and let people see who you are," he said.

"No, I don't want to," Hatfill replied, according to Clawson.

"Well, you're going to have to. You're getting eaten alive."

"Look, you know how my lawyer feels about it," Hatfill replied.

Soon after, Clawson spoke to Victor Glassberg, Hatfill's lawyer. "We had a tough conversation," Clawson said, referring to their opposing views about the need for a press conference. But on August 11, in front of his lawyer's office in Alexandria, Virginia, Steven Hatfill proclaimed his innocence in a statement to the press. This was followed, Clawson said, by a "disinformation campaign" by the FBI, including the bureau's denial that they had trashed his apartment. "But we had pictures," Clawson said, "which we released to the press." On August 25, Hatfill held another press conference.

Telecast nationwide, a resolute Steven Hatfill pointed his index finger toward the assembled cameras and said: "I want to look my fellow Americans directly in the eye and declare to them, 'I am not the anthrax killer.'" His dark blue suit covered the stocky frame of a seemingly over-age wrestler. He inveighed against Barbara Rosenberg and Nicholas Kristof for conveying a "never-ending torrent of leaks." He assailed Attorney General John Ashcroft for singling him out as a "person of interest." And he accused the government of abusing him: "This assassination of my character appears to be part of a government-run effort to show the American people that it is proceeding vigorously and successfully with the anthrax investigation."

Hatfill provided timesheets to reporters showing that he had been working overtime at the SAIC offices in McLean, Virginia, on September 17 and 18 and October 8 and 9, around the times the anthrax letters were mailed. "I know nothing about the anthrax attack," he said. "I had absolutely nothing to do with this terrible crime."

After the second press conference, Hatfill moved from Frederick, Maryland, to his girlfriend's apartment in Washington, D.C. But the FBI remained interested in his former home and the surrounding area. At the time that I spoke with Clawson in January 2003, investigators were searching a wooded area near Hatfill's former home. The search was "just a continuation of our investigation on the anthrax case," according to FBI spokeswoman Debra Weierman. In June the FBI returned to the area and drained a 1-acre pond at a cost of $250,000.

I asked Clawson how Hatfill was spending his time. "He watches CNN a lot," Clawson replied. "He is angry that his reputation is in tatters and that he has been reduced to virtual poverty."

Randy Murch put his cup of coffee on the saucer, removed his black suit jacket, and draped it over the back of his chair. He tilted his head forward. We were in a corner of the breakfast room of a hotel in northwest Washington. "I coined the term 'microbial forensics' when I was with the FBI," he said. After 23 years with the bureau, Murch, 50, retired at the end of 2002 to work on terrorism issues at the Institute for Defense Analyses, a think tank in Alexandria, Virginia. Now, 2 months later, in February 2003, he discussed his former role as deputy assistant director in charge of forensic programs in the FBI's laboratory. He lifted a pair of gold-rimmed reading glasses from his pocket, slipped them on, and sketched a diagram of the bureau's hierarchy. His position was on a rung parallel to that of the head of counterterrorism, Robert Blitzer, with whom he often worked. They were below the FBI director's slot, separated from the director by only the deputy director.

Murch spent much of his career on the third floor of the FBI's Washington headquarters building. On that floor, laboratories are replete with chemical reagents, incubators, centrifuges, and microscopes. "You see the same instruments and skill sets here as at other labs," Murch said. But an outsider is also struck by a difference. Shelves and tables are filled with items that seem eerily out of place: a dress, a shirt, a smashed headlight. Here a knife, a revolver, there a blood sample, a urine sample. "Substrates," Murch called them—items that will be analyzed for fingerprints, DNA typing, illicit chemicals, or any such possible evidence.

After receiving a Ph.D. in plant pathology from the University of Illinois in 1979, Murch began working for the FBI. Almost from the outset he was worried about biological terrorism, though few others at the time were contemplating the subject. But in 1996, with increasing concern about terrorism in general, he was able to establish the bureau's Hazardous Materials Response Unit. "Just in time for the rash of bioterrorism hoaxes that began the following year," he quipped. Murch spoke softly through a closely clipped mustache and beard. In the period after September 11, 2001, his efforts largely focused on suspects and collaborators connected with the attacks. Through court-ordered electronic surveillance he was locating, tagging, and tracking them.

What did he think of the FBI's profiling of the likely anthrax mailer? I asked. Murch answered:

> You know, any investigation of this sort is not just science. It's part art, part experience, part luck. I wasn't that deep in the anthrax investigation per se, but what the counterterrorism division and the laboratory division did was synthesize the information and forensic evidence available. And they probably looked at earlier experiences with similar kinds of activities. I know they consulted with outside scientists to develop their approaches.

Did he think the perpetrator was domestic? "I think it was either of two possibilities: a homegrown guy who had the stuff on a shelf and took advantage of the September 11 timing or, maybe, somebody from outside who sent it to a local operator."

Murch is inclined toward the former possibility. Whoever did it, he said, knew his way around the United States and the East Coast in particular. "It may not be Hatfill, but I think it is somebody like him." Murch tilted his head back and said that he predicted years ago that some domestic person would use biological agents as terror weapons. "You don't need much equipment or an advanced degree to make biological weapons," he said. "You could fit all the stuff in a garage."

Murch's interest in microbial forensics has brought him in touch with many outside scientists through the years. A decade earlier he met Paul Keim and Paul Jackson, who were working on DNA fingerprinting at the Los Alamos National Laboratory. He sensed then that what they were doing could be important to criminal investigations. In the wake of the anthrax attacks, the FBI did indeed turn to Keim and others in the scientific community for help.

On October 5, 2001, the day that Bob Stevens died, the *Journal of Clinical Microbiology* accepted a paper titled "Molecular Investigation of the Aum Shinrikyo Anthrax Release in Kameido, Japan" for publication in its December issue. Aum Shinrikyo was the cult that in 1995 released sarin nerve agent in the Tokyo subway. In 1993 the same cult disseminated anthrax bacteria from the roof of a facility in Kameido, a suburb of Tokyo. Unlike the nerve agent, which killed 12 people and injured more than a thousand, the anthrax had no discernible effect. The reason had been unclear. Now this new study would confirm that the cult's anthrax had been the Sterne strain, a nonpathogenic bug used to vaccinate animals.

A decade earlier Paul Keim, principal investigator of the study, was struggling to stretch his $50,000 grant to run his laboratory at Northern Arizona University. Things began to change after he developed a technique to differentiate strains of anthrax. By the beginning of October 2001 his grants had climbed to $1.6 million. Before the end of the month the figure had doubled to $3.2 million. The FBI and other government agencies were pouring money into Keim's laboratory with the hope that he could help identify the source of the bacteria in the anthrax letters. In fact, it was Keim who confirmed that the anthrax bacteria in the letters were of the Ames strain, the same strain as in the infected patients.

Keim received a Ph.D. in 1981 from the University of Kansas, followed by postgraduate work at the University of Utah. Despite his new celebrity, Keim's attire still consists of blue jeans, running shoes, and, when temperature permits, a T-shirt. His lanky frame and long face are topped with a full crop of blond hair. When I told him that he seems younger than his age, 46, he chuckled, "Let's see. What do they call me in the press? Oh yes, a young civilian scientist." He ended the sentence in full-throated laughter.

Keim has amassed 1,350 anthrax strains from around the world, the largest collection anywhere. "Actually, whether they are called strains or isolates is hotly debated in the microbiology community," he said. He added that the more we learn about the genetic makeup of bacteria, the more challenging it is to classify them. "Bacteria are exchanging DNA all the time," Keim explained, "and some species do so more frequently than others."

Escherichia coli, a common intestinal bacterium, is a good example of the variability problem. To survive and thrive, these bacteria must divide often, producing around 300 generations per year. This frequency of division increases the chances for mutations—alterations of their genes—which explains the large diversity in the genomic structures of *E. coli*. "Two different strains of *E. coli* have been entirely sequenced," Keim said, "and they only share about 60 percent of their genes in common."

Bacillus anthracis is very different. Anthrax spores can lie inactive in the soil for a century or longer. During that period the chances for DNA exchanges are nil. If spores happen to infect a cow or other livestock, they will germinate, reproduce, and become open to the possibility of mutation. But because anthrax outbreaks are infrequent, the long-dormant bacteria remain almost identical to each other.

Before Keim's innovation, differentiating between strains of anthrax bacteria was very difficult, and identifying distinctions within one strain was almost impossible. But in the mid-1990s Keim began collaborating with Paul Jackson, a colleague who worked at the Los Alamos National Laboratory, to scan the anthrax genome. Using specialized "restriction" enzymes, they cut DNA strands at specific sites and broke up the anthrax genome (which is composed of 5.5 million base pairs) into fragments. Then, Keim explained: "We'd grab restriction fragments out of the genome and compare those same restriction fragments across different strains and see if we could find differences. We looked at thousands. The fragments were pretty short, maybe 250 base pairs."

In this way, Keim could detect change even if by only a single nucleotide, or DNA base. And this provided the ability to discover the rare genetic inconsistencies even within a single strain. "So we could take about 100 isolates of *Bacillus anthracis* and look at a thousand or more of these restriction fragments. And when we did that, we found only 30 fragments that varied among the strains." The tiny variation impelled Keim to claim that "*Bacillus anthracis* is the most homogeneous bacterium in existence."

Keim's technique is called MLVA, for Multi-Locus VNTR Analysis. The VNTR stands for Variable Number Tandem Repeat. As he heard himself speak the double mouthful, he chuckled, "I've been blamed for creating an acronym within an acronym." So what exactly is it that Keim's MLVA does that compares anthrax bacteria with each other? "OK. What we've done is identify the most informative regions of the *Bacillus anthracis* genome for discriminating among isolates."

What does he mean by "informative regions"? He responded by way of analogy:

> If all humans were 6 feet tall, you get no information from that characteristic that would set them apart. But if you have a criterion to look at those 6-foot-tall people, say, by four or five different hair colors, that provides information for telling them apart. So you've got to find a characteristic of whatever you're looking at, that varies.

In February 2002, Keim announced at a news conference that he had found distinguishing features among stocks of Ames anthrax that were stored in different laboratories. His findings seemingly would allow for tracing the mailed anthrax to one of the laboratories through his MLVA technique. But on instruction from the FBI, for whom he was doing these assessments, he refused to name the laboratories from which he obtained the bacteria.

A year later, no publicly known information had yet tied the anthrax in the letters to a particular laboratory. I asked if Keim still thought it possible to identify the laboratory of origin. Again citing his agreement with the FBI, he spoke cautiously: "I really can't answer that other than to say it's possible."

As the investigation of the anthrax letters unfolded, Keim himself learned a lot about the source of some of the strains in his collection. When we spoke, early in 2003, he said, "You know, 18 months ago, I didn't know the Ames strain had come from Texas or when it was discovered." The strain was first acquired by the army for vaccine research at Fort Detrick, Maryland, after it infected some Texas cows in 1981. The bacteria were so virulent that samples were later sent to other military and civilian laboratories for study. But an early shipment used the return-address label of a government laboratory in Ames, Iowa. Confusion over the strain's origin arose in 1985 after a scientific paper mistakenly assumed that the return-address was the original source of the bugs.

The U.S. germ arsenal was destroyed after President Nixon's decision in 1969 to eliminate this country's offensive biological weapons program. At that time the Army's anthrax weapons had been made up of the Vollum strain. Thus, as Keim belatedly learned, the Ames bacteria could not have been part of the Army's earlier biological arsenal. Although the Ames strain was found in the United States, samples had been sent to several laboratories overseas, including the British biological defense establishment at Porton Down. Which leads to the question of whether the source of the material in the anthrax letters might have been outside the United States. "That's a really interesting question," Keim said. "But I have totally disengaged my mind from that kind of speculation." He sees his job as purely scientific. "My role in the investigation is very focused. We look at DNA. We try to identify. We try to discriminate. For me to speculate beyond that would be a distraction from what I'm doing."

On November 21, 2002, the Perry Videx Company in Hainesport, New Jersey, hosted an unusual group of visitors. The company sells equipment that is used by pharmaceutical, chemical, and food manufacturers to process their ingredients. But the 20 visitors were not interested in the traditional applications. Rather they wanted to know if the equipment could be applied to more nefarious purposes, such as the development of chemical or biological weapons. The group was learning what to look for in preparation for going to Iraq. They were United Nations weapons inspectors, members of UNMOVIC, the United Nations Monitoring, Verification and Inspection Commission.

U.N. Security Council Resolution 1441 had just been enacted on November 7. Like other resolutions since the 1991 Persian Gulf War, it noted Iraq's failure to provide "full, final, and complete disclosure" about its weapons of mass destruction. Now, under military threat from the United States, the Iraqi regime was being offered "a final opportunity to comply with its disarmament obligations," according to the resolution. It would be the inspectors' task to report on Iraqi compliance.

Gregg Epstein, president of Perry Videx, walked the group through a cavernous warehouse filled with bulky equipment. He paused at the mouth of a huge steel cylinder positioned horizontally. Its thick door was swung open, as if beckoning entry into a bank vault. Epstein explained that the piece is a centrifuge that separates large quantities of liquids from solids. "This one was actually used to make vitamin C," he said, "but bad stuff like sarin or VX nerve agents would corrode the metal." He pointed at the metal inner wall and said that if it had been lined with a corrosion-resistant material, like Teflon, the inspectors should be suspicious.

He moved to a 6-foot tall metallic cylinder about 3 feet in diameter, capped with a removable dome. A half dozen pipes sprouted from the side and base. "This is a spray dryer," he said. Manufactured by the Niro Company, it is used in pharmaceutical and food processing to make free-flowing particles that can be microscopic in size. A slurry of material and a fine powder are fed in through the top. From the side a pipe conveys hot air into the cylinder while another one draws cooler air away. Although spray dryers are used in the manufacture of a variety of innocuous products like toothpaste and flavored foods, they can also be used to produce biological warfare agents.

It is just such a piece of equipment that Richard Spertzel has in mind when he thinks about the possible source of the anthrax letters. Contrary to the FBI and many other observers, Spertzel finds the "lone wolf" theory unlikely. In December 2001 he told the House Committee on International Relations that descriptions of the anthrax in the Daschle letter indicated "it could be produced only by some group that was involved with a current or former state program in recent years."

Spertzel's favored candidate for the source was Iraq, where he had previously served as a weapons inspector for the United Nations Special Commission. The Daschle letter contained anthrax that was more pure and concentrated than any that had been found in the Soviet, U.S., or Iraqi biological programs. But, Spertzel noted in his testimony, "Iraq, unlike the Soviet and U.S. programs, did not mill its dried product; rather the Iraqi BW [biological weapons] team learned the method of obtaining readily aerosolizable small-particle product in a one-step spray-drying process."

A year later, in September 2002, Spertzel testified before the House Armed Services Committee:

> Although Iraq denies it, Iraq had the equipment and know-how to dry BW agents in a small particle that would be highly dispersible

> into an aerosol. . . . It still retains the necessary personnel, equipment (including spray dryer), and supplies to have an equal or expanded capability in this regard.

Few can match the experience Spertzel brings to the discussion. After graduating in 1960 from the University of Pennsylvania School of Veterinary Medicine, he began a 28-year Army career. While in the service, he received a Ph.D. in microbiology from Notre Dame. But most of his career was spent at Fort Detrick where, until the end of the U.S. offensive biological arms program in 1969, he helped develop germ weapons, including anthrax.

Spertzel cited his expertise to underscore how unlikely the lone wolf theory is. "In my opinion," he told the *Washington Post* in October 2002, "there are maybe four or five people in the whole country who might be able to make this stuff, and I'm one of them." He was referring to the unusually pure mix of the anthrax in the Daschle and Leahy letters. "And even with a good lab and staff to help run it, it might take me a year to come up with a product as good." (Others believe that developing the product would be easier than Spertzel suggests. Matthew Meselson and Ken Alibek say that the micrographs they saw of spores in the Daschle letter showed no added materials. Each of them told me he thought the spores could have been prepared by any skillful microbiologist.)

In 1994, Spertzel began a 4-year tour with UNSCOM, the predecessor of UNMOVIC. Serving as the commission's head biological weapons inspector, he oversaw UNSCOM's finding in 1995 that Iraq's germ weapons program had been far more advanced than the regime previously acknowledged. Moreover, the commission determined that Iraq was still hiding information. Spertzel himself had found "a major disparity between the amount of agent declared as produced by Iraq and that estimated by UNCSOM experts," he later told Congress.

Paradoxically, despite the evidence, several members of the U.N. Security Council had begun to criticize UNSCOM for being too aggressive. By 1998, Russia, France, and China even seemed to be challenging the integrity of the inspectors. Spertzel had decided the commission had become a political football and could no longer function usefully. He left in July. In December, President Clinton ordered the bombing of Iraqi installations with the proclaimed aim of degrading Iraq's weapons of mass destruction. Subsequently, Saddam Hussein prohibited the return of UN inspectors. The next year UNSCOM was disbanded and replaced in December 1999 with a newly staffed UNMOVIC.

Richard Spertzel's voice is soft. His nasal twang hints at his western Pennsylvania origins, where he grew up on a farm. In a dark checked sports jacket, he squinted through metal-framed glasses as he emphasized a point. Despite his quiet demeanor, not long into a conversation it becomes clear that he is a man of forceful convictions.

When I asked Spertzel to comment on the anthrax mailer scenario I posed at the beginning of this chapter, his observations were similar to Barbara Rosenberg's. Despite their differences about the source of the anthrax letters, they both found the depiction largely plausible, at least for the letters to NBC and the *New York Post*. Like Rosenberg, Spertzel allowed that the anthrax in those letters could have been placed in the envelopes as I described, though not the material later sent to the two senators. Spertzel:

> The quality of the product in the Daschle and Leahy letters is such that it could not have been 'tapped' into any receptacle. It would almost all become airborne. Similarly it could not have been transferred by 'sifting' into a plastic test tube. After it was in a sealed container it would take many hours to settle because of its lightness and flowability.

Exactly how that finely graded anthrax was transferred to the envelopes remains puzzling. Perhaps a sucking mechanism of some sort was used to draw the powder into the envelopes.

I asked Spertzel to talk further about weaponizing anthrax while not, of course, giving an actual recipe (not that he would in any case). Spertzel began: "You know, the pharmaceutical industry uses a number of silica compounds in the manufacturing of inhalant medications." Silica, or silicon dioxide, is a hard glassy mineral that forms a variety of familiar substances, including sand and quartz. In microscopic size it can adhere to the surface of other particles and keep them from sticking to each other. It is the silica-coated particles that enhance the slippery characteristics of products like paint and toothpaste. Spertzel noted that since the 1950s silica compounds could also be added to biological agents "which makes them 'flowable' and readily dispersible in the air." (The coating reduces the electrostatic attraction that the agents might otherwise exhibit.) He mentioned Aerosil, the commercial name for one such compound. It is a fine white powder whose particles are about 12 nanometers, a thousandth the size of an anthrax spore. He explained further:

> To get the kind of product that was in the Leahy and Daschle letters,

> the silica would have to be added before the anthrax was dried. There's only one drying technique I know that will give you that narrow range of small particle aerosol. And that's spray drying.

That does not sound very complicated, I said. "Well, it's not super complicated, but it's a very exacting operation to get to the desired particle size. You need a lot of trial and error," Spertzel said.

Routine spray drying might produce a batch of particles, some of them 20 microns in size or larger, some perhaps smaller. The larger particles are not individual spores but a collection of several spores that have stuck to each other during the processing. They emerge from the spray dryer as a single particle whose surface is coated with silica. So how could someone produce a "pure" mix of individual free-floating spores?

> You'd need a spray dryer that is called a "co-current" dryer. One of the currents is heated air and the other current is the material you want to dry. And the exact proportion of these two streams coming together is what will determine the particle size. Increasing the heated air relative to the material you're drying would give you a smaller particle.

Spertzel cannot fathom any other way that the process could have yielded silica-coated particles of 1 to 3 microns, scarcely larger than the naked spore. The process, he reiterated, would require repeated adjustments to the heated air flow. And after each trial run, an examination of particle size would be needed, perhaps under a scanning electron microscope. "Trial and error, trial and error," he repeated.

A small room would be inadequate to process the anthrax, Spertzel insisted, as he tallied up the necessary equipment: refrigerator, incubator, spray dryer, maybe a 3-foot-wide scanning electron microscope. And all the work would have to be done in a thoroughly contained area. "I would say the space you need would be something like 20 × 50 feet. This is why I say it's got to be something that was done with the complicity of the country in which it was made," he concluded.

Assuming the complexity, time, and space that Spertzel specified, production of the finely graded anthrax would seem unlikely to have taken place in a known U.S. laboratory. There would be too many people around for the effort to go unnoticed. On the other hand, constructing a "safe" room in some obscure location seems less improbable. Thus, to flatly rule out the possibility of a domestic loner, as Spertzel does, seems an overreach. Still, one can hardly deem his conjectures unreasonable.

He noted that documents uncovered by UNSCOM showed that Iraqi agencies had sought to obtain the Ames strain in 1988 and 1989. There were no restrictions at that time to obtaining pathogens from laboratories around the world. He thinks a likely source could have been the Pasteur Institute in Paris, which had several strains. "We know the Iraqis obtained many strains of anthrax from the Pasteur Institute," he said, "but we only know what two of them were." (In March 2003, I inquired of Michèle Mock, director of the institute's Annual Report on Toxins and Bacterial Pathogenesis, if the Ames strain had ever been stocked there. Her e-mail reply: "No, we never had the Ames strain at Pasteur.")

How would all this relate to September 11 and Osama bin Laden?

Spertzel said there is evidence that demonstrates a connection between Iraq and Al Qaeda. He mentioned that three Iraqi defectors who defected at different times "all told the same story" about Iraqi cooperation with Al Qaeda, and "I have personally spoken to two of them." Convinced of their credibility, it is no great leap for Spertzel to believe that there was collusion between Iraqi operatives and the people responsible for September 11. Around the time of our conversation, in early 2003, members of the Bush administration also were emphasizing that there was a linkage. In February, Secretary of State Colin Powell provided the U.N. Security Council with evidence that Al Qaeda operatives had been working with the Iraqis. Whether this meant collusion concerning the anthrax letters remains uncertain.

When I reminded Spertzel that many people disagree with his views about the source of the anthrax letters, he replied, "I know they do. And that's the reason I'm willing to sit back and wait for a couple of years and say, 'Look you bastards, I told you so.'" With a wink and a laugh he added, "If I live long enough."

"I am still always struck by how disturbingly 'normal' most terrorists seem when one actually sits down and talks to them," Bruce Hoffman wrote in his 1998 book, *Inside Terrorism*.

Simply put, a single terrorist "personality" does not exist, which can compound the difficulties in identifying and finding a terrorist. But if personality type is elusive, patterns of motivation seem more accessible. As Hoffman suggests, the defining characteristic of terrorism is the quest for a political goal through violence or the threat of violence. The term "political" here has religious as well as secular connotations. A terrorist may use the same instruments as those of a lunatic killer—guns and bombs—but his motive is different. John Hinckley's attempt to murder President Reagan in 1981 was not an act of terrorism. Rather, it was a perverted effort to impress the actress Jodie Foster. It was grounded in a wish for personal aggrandizement, not prompted by politics or ideology.

In contrast to Hinckley-type motivation, the terrorist is an "altruist," Hoffman wrote. His action is aimed at achieving a greater good, not personal benefit. The cause may be overarching, such as remaking society, or it may be narrow, as in opposing nuclear power or abortion (a reminder, again, of anthrax hoaxer Clayton Lee Waagner).

During a conversation in early 2003, Hoffman expressed an elevated sense of frustration. He is the director of the RAND Corporation in Washington, and we met in his 8th floor office above Pentagon City Mall. In corduroy pants and shirtsleeves, Hoffman was comfortably informal as he responded to my question about who might have been behind the anthrax letters: "I have to say that after September 11 any terrorist analyst who ventures a call on this is on thin ice. September 11 wiped the slate clean of our assumptions. To me, on all aspects of the anthrax case, I'm completely agnostic."

He was alarmed by the quality of the anthrax and the unanticipated effects it had when sent in the mail. "Before October 2001, we never thought that inhalation anthrax would be caused by the mail. Cutaneous maybe, but not inhalation." He pondered as well the unexpected nature of the September 11 attacks—simultaneous hijackings by suicide terrorists. Those events, he said, undermined old assumptions about who might be a terrorist and how they would conduct their acts. It is clear now, if it had not been before, "that a lone individual can be a terrorist, not part of a group, though he might be animated by a political cause." But Hoffman went no further and refused to offer an opinion about who was behind the anthrax letters.

Other terrorism experts feel less constrained, including Jessica Stern, a professor of public policy at Harvard. She has examined the motives of terrorists who would commit acts of "extreme violence," such as the use of biological, chemical, or nuclear weapons. These "ultimate terrorists," as she calls them, are increasing in number. They have sidestepped the dictum made famous in the past by another terrorism expert, Brian Jenkins, who said that terrorists want "a lot of people watching, not a lot of people dead." For Stern the new type of terrorist is motivated by religious conviction or violent right-wing ideology. Prompted in some instances by desire for revenge, she believes, these individuals are more likely to use weapons of mass destruction.

In fact, Timothy McVeigh and the acolytes of Osama bin Laden nicely represent Stern's two types of terrorists that revel in mass casualties. McVeigh, the right-wing extremist and bin Laden, the religious fanatic, both engineered attacks on core U.S. institutions. In 1995 McVeigh exploded a truck bomb at the Alfred P. Murrah Federal Building in Oklahoma City, killing 168 people. In 1993 Islamic militants bombed the World Trade Center, causing the death of six. And in 2001 the bin Laden-inspired attack on the World Trade Center killed nearly 3,000. During the past decade many more terrorist acts could be traced to radical Islamic sources than to any other.

A year after the anthrax attacks, I asked Jessica Stern who she thought was responsible for the letters.

> From early on I've thought that the most likely person was sympathetic to the American right-wing extremists. But that was based just on intuition. I think of these movements as an influence on individuals who may act as lone wolf avengers even though they are not actual members of a group.

Stern does not rule out the possibility of an Al Qaeda connection, but she is doubtful because "nobody has provided us evidence that Al Qaeda was involved." Stern's suspicions are strongly influenced by her own interview experiences with right-wing extremists.

> My inclination is based on what they've said to me. The guy who heads the Montana Militia, John Trochman, told me, "Biological weapons are in the air in the American right-wing [extremist] movements." He meant that it's just something they think about all the time.

Another expert, Jerrold Post, brings to the discussion the perspective of a psychiatrist who has studied the motivations of terrorists. He agrees that they represent a "wide range of psychologies," including many that seem normal. But he has the impression that a disproportionate number are narcissistic sociopaths who are self-absorbed and have low tolerance for frustration. Post, who directs the Political Psychology Program at George Washington University, has a deep, gravely voice. "I'm probably the world's leading terrorist interviewer," he said matter-of-factly. He has interviewed 35 incarcerated Palestinian terrorists from Hamas, Islamic Jihad, and Hezbollah. And he has testified at several U.S. federal trials involving terrorism. When I asked his thoughts about the possible anthrax mailer, he began with a general assessment.

He believes that most terrorists feel constrained from using weapons of mass destruction. "For the large majority of terrorists, the goal of their act is to call attention to their cause and to win positive attention." (Sounds like Brian Jenkins.) Many of the terrorists he interviewed said, in effect, "Just give me a good Kalashnikov assault rifle." They think it might be nice to have a weapon that could kill 10,000, Post said, but the idea is also scary to them. He recalled that a religious terrorist had told him the Koran proscribes poisoning and that "it would be against our religious belief to get involved with that."

Post thinks that two types of terrorists are less constrained about the use of weapons of mass destruction such as (potentially) anthrax: religious extremists and the "lone right-wing scientist with a gripe." Therefore, he said, it is unlikely that Iraq was connected to the anthrax letters. Why not Iraq? Post explained:

> Even though Saddam Hussein had developed his own anthrax weapons, in terms of my own understanding of his psychology, he is a very prudent individual. He surely knows that if it could be traced back to him, he'd be incinerated. I don't see him ever letting such materials out of his hands unless it's with a totally controlled group.

Then who *would* try to terrorize by sending the anthrax letters? Post suspects that the motive might have been personal, perhaps a quest for revenge. He thinks the anthrax mailer had expertise in microbiology and harbored "some twisted places in his or her psychology." Further, the mailer "probably has some dreams of glory, had not been achieving very much in life, but has gained profound satisfaction from what his few letters have accomplished." Post's description fits the FBI's profile of the lone perpetrator.

A former high-level FBI official, an expert on counterterrorism, also thinks the disaffected loner theory is probably correct. I asked how he feels about Steven Hatfill. "He fits the bureau's profile," the former agent acknowledges, "but I don't think he's the guy." Requesting anonymity, the ex-agent, who now works for a government defense contractor, says that he knows Hatfill. "I've traveled with him and I've socialized with him through work."

After allowing that Hatfill impressed him as being an odd man, he described Hatfill as very patriotic. "Clearly, he knows the danger of mailing anthrax, but he doesn't seem to me to be someone who could kill somebody." The agent's assessment stems from his long experience in criminal investigations.

What does the agent think of the unusual speed with which the FBI came up with its profile? He answered obliquely:

> I've never been a big fan of profiling. Look at the two Washington-area snipers who were apprehended in October 2002. The profilers had it as a white male. [It turned out to be two black men.] You can't base your whole investigative strategy on a profile. It's just not always right.

So who does this veteran counterterrorism expert think is responsible for the anthrax attacks? "I think the likelihood is in the following order. First, a loner. Second, Al Qaeda, in conjunction with the September 11 attack. Third, Iraq or another government."

The former agent returned to Hatfill, expressing sympathy for his current situation. "I think Hatfill is unemployable now," he said after alluding to the publicity given to the FBI's investigation of him. The ex-agent does not exclude the possibility that Hatfill is guilty. "Look, he could have done it. He knows the scientific stuff inside and out. The question is, 'Did he do it?'" "Basically, I'm 70 percent that he didn't do it. But 30 percent that he might have," said this counterrorism expert. He thinks the matter will be resolved some day but not necessarily soon. "Cases like this one can go on for years without being solved. Then a single break can come."

In mid-2002, FBI director Robert Mueller denied that the bureau had ruled out any group or laboratory as the source of the anthrax letters. Still, a year later the FBI Web site on Amerithrax continued to display the profile it had posted back in November 2001. The focus remained on a probable domestic perpetrator—a nonconfrontational adult male who "prefers being by himself more often than not." Is this tilt toward the lone operator, which has been endorsed by many others, warranted? Perhaps eventually it will prove to have been. Meanwhile, it would seem irresponsible to ignore the possibility of an overseas connection to the anthrax letters.

Dr. Larry Bush, the physician who diagnosed Bob Stevens's anthrax, shakes his head in disbelief about the lone perpetrator theory. "For one thing," he asked, "whoever heard of American Media?" He was referring to the company that publishes the *National Enquirer* and the *Sun*, which were targeted with anthrax. "I live in the neighborhood, and I didn't know who published the *National Enquirer*. Who would know that?" he asked again. He ticks off some familiar coincidences: that several highjackers had lived a few miles from the American Media building and that the wife of the editor of the *Sun* had rented apartments to two of them.

Then there is the fact that in June 2001 one of the highjackers was treated at a Miami hospital for a black lesion on his leg. In hindsight the hijacker's physician believed it was cutaneous anthrax. Finally, there was the extraordinarily brief period, only 6 days, between September 11 and the day the first letters were mailed. Bush offered a half smile. His expression was the same as the one he wore when telling me how people initially dismissed his diagnosis of anthrax. It was a wordless reminder that his suspicion turned out to be right after all. Then he said: "When I look at all this stuff about the hijackers, I say, 'Well, wait a minute. This sure looks suspicious.'"

The notion of a connection between the September 11 terror and the anthrax terror remains speculative. But so is the case for a lone domestic perpetrator. Given the information that is publicly available, either maddening scenario seems possible.

For all the profiling and theorizing, it is quite possible that the anthrax mailer will never be found. In November 2001 the *Record*, a New Jersey newspaper, indicated that state authorities had investigated 178 reports of anthrax in the 3 years before September 11. All turned out to be hoaxes, and not a single culprit was ever identified.

The question of who was behind the anthrax letters is tantalizing and important. But it is only one of many unresolved matters related to the anthrax letters. The issues range from dealing with buildings that are still contaminated with spores to preparing for future bioterrorism, and the preeminent challenge of all—how to minimize the chances that anthrax or other biological weapons will be used.

CHAPTER
TEN

Loose Ends

CONTAMINATION

The signs finally came down in 1990. No longer were the graceful green slopes of Gruinard Island, off the northwest coast of Scotland, forbidden land. Since World War II, when government biological warfare experiments were conducted there, bold red lettering around the island had ominously warned: The Ground Is Contaminated With Anthrax and Dangerous. Landing is Prohibited. Now, nearly half a century later, after a monumental cleanup effort, the Ministry of Defense deemed the island safe again.

Until the early 1980s no one was sure whether the island would ever be decontaminated. Its 522 acres seemed too large an area to cover. Then, careful testing indicated that the spores were concentrated in a 5-acre area, in the top 3 inches of soil. That fortunate finding coincided with mounting pressure for action by worried mainlanders and members of Parliament, and the British government decided to try to restore the island to a safe condition. The challenge was formidable given the absence of experience in decontaminating any comparable area. But a scientific advisory group to the government surmised that a properly distributed solution of formaldehyde could destroy the bacteria.

The presumption was based on an earlier experience with an anthrax-contaminated building in Manchester, New Hampshire. In 1957 the Arms Textile Mill along the Merrimack River received a shipment of anthrax-laced goat hair from Pakistan. The discovery of the tainted shipment was made after nine employees came down with the disease. Four of the five who became infected with inhalation anthrax died. Coincidentally, Dr. Philip Brachman, who had been leading a CDC (then the Communicable Disease Center) investigation of an experimental anthrax vaccine at that plant, found that none of the immunized workers had become ill. Vaccination then became required of all employees until the building was closed in 1968. Fear about spores there continued to linger, however. A 1970 headline in the *Manchester Union Leader* declared "Lethal Spores Could Menace All Manchester." Soon after, a decision was made to decontaminate the structure.

In 1971 workers in oxygen masks and protective gear walked through the building spraying detergent, and then formaldehyde vapor, in every direction. Five years later, as the building was being demolished, it was sprayed with a chlorine solution. The wooden remnants were incinerated, and the bricks were carted a half-mile away for burial deep under the earth.

The decision to use formaldehyde had been based on its well-established germicidal qualities. Its sweet pungent odor was familiar in operating rooms and other health care facilities. The mill in Manchester had been the largest decontamination project until that time. Then came the vastly greater challenge of Gruinard Island.

In 1986 a multi-step approach to cleanse the five contaminated acres of Gruinard Island was begun. First came spraying with a herbicide and a burning off of the dead vegetation. Then an elaborate network of perforated tubing was snaked across the denuded ground surface. From June through August a 5 percent concentration of formaldehyde in seawater was pumped through the tubes to cover one land segment after another. An estimated 50 liters slowly soaked into every square meter of soil (that is, about a dozen gallons of formaldehyde mix per square yard).

Months later soil samples showed that most, but not all, of the affected area was free of viable anthrax spores. The tainted locations were retreated in July 1987. After further tests, in 1988 the science advisory group declared the chance of contracting anthrax on Gruinard Island was remote. By then the island had been drenched with 280 tons of formaldehyde diluted in 2,000 tons of seawater. The government said in 1990 that the island was safe, and the restrictive signs were removed.

Despite the assurance, few people have set foot on Gruinard Island. In 2001 a British *Telegraph* reporter asked a middle-age woman in the village of Laide, across the bay, about her reluctance to go there. "You never know, do you?" she replied. A man in a nearby shop had a similar reaction. "Why take a chance?" he asked. "Life out here is hard enough as it is."

Gruinard Island exemplifies the challenge of addressing widespread anthrax contamination, even though formaldehyde is known to kill spores. The chemical has been used to decontaminate locations in the United States as well, notably Building 470 at Fort Detrick, Maryland. Known as the "Tower," it is a seven-story brick structure in which anthrax and other biological warfare agents were produced. Since 1969, when the United States ended its offensive biological weapons program, the building has gone unused. Large steel tanks, in which thousands of gallons of anthrax slurry were once mixed, stand gray with dirt and age. The building has been gassed with formaldehyde vapor, but suspicion remains that spores might be buried in the crevices. An Army publication indicated in 1993 that despite "three successful decontamination procedures [Building 470] was not certified 100 percent clean." Government officials contend the building poses no danger and that plans are under way to have it razed.

Given the experiences with Gruinard and the Tower, concerns about contamination wrought by the anthrax letters were not surprising. Spores were found in scores of locations along the paths of the letters. Where contamination was discovered in limited sections of buildings and post offices, cleanup was accomplished in a few months. But large structures, including the American Media building in Florida and the postal centers in Hamilton, New Jersey, and Washington, D.C., remain sealed and off limits.

Tom Day is a round man with a friendly smile, a beardless Santa Claus. A 20-year veteran of the U.S. Postal Service, he has been vice president for engineering since 2000. One of his responsibilities has been to oversee the cleanup of anthrax-contaminated post offices. Spores were found in 23 mail facilities, but only the sorting centers in Hamilton and Washington were infested throughout. "The other 21 were cleaned up by surface scrubbing with bleach," he said. "Post-testing assured us that it was effective."

A West Point graduate, Day grew up on Long Island where he became a devoted New York Yankees fan. He named the players to me in the large painting on his office wall in Merrifield, Virginia: "That's Thurmond Munson tagging out Steve Garvey in the 1978 World Series." He laughed, and added that the Yankees beat the Los Angeles Dodgers in six games.

We sat at a small brown table 20 feet from his desk. It was early November 2002, a year after the anthrax letters left their trails of spores. Cost projections for remediation had already tripled beyond earlier estimates. In March the postal service had figured that cleaning up the Brentwood center would come to $23 million and Hamilton $13 million. Day's voice turned lower: "Now it looks as though both of them together will cost more than $100 million." In fact the meter was still running. Two months after we spoke, projections had risen to $100 million to clean up Brentwood alone and another $50 million for Hamilton. The estimates were based in part on the experience with the cleanup on Capitol Hill, including the Hart Senate Office Building. The cost there ended up at $41.7 million, nearly double the initial projection of $23 million.

"We assisted the Environmental Protection Agency in cleaning up the Hart Building and we learned a lot from that experience," Day said. There, chlorine dioxide was used for the first time in a large-scale decontamination. Why not formaldehyde? I asked. Day said that they considered using formaldehyde and several other antimicrobial agents, including methyl bromide and ethylene oxide, but they're all quite toxic, carcinogenic. On the other hand, chlorine dioxide, which has long been used to kill germs as part of the purification process in water treatment plants, "quickly decays and becomes a harmless substance."

The chlorine dioxide treatment in the Hart Building was not an immediate success. After an initial spraying in December, viable anthrax spores were still found. A second fumigation effort ended with similar results. Finally, a third release of the chemical apparently killed the remaining bacilli, and the building was reopened in January 2002.

Decontamination of the postal centers posed a much bigger challenge. The treated section of the Hart Building measured 3,000 square feet. The Brentwood facility, at 632,000 square feet, was 210 times larger. Planning and bleach scrubbing in advance of the chlorine dioxide release at Brentwood were not completed until December 2002. In the third week of December, 200 workers assembled at Brentwood amid 30-foot-high chemical tanks, diesel-powered generators, and 21 miles of intricately laid-out plastic tubing. Humidity in the building had been raised to 70 percent and the temperature to 75 degrees. These conditions would optimize the effects of the gas. Then, 60 tank trucks arrived, each with a police escort. They carried 270,000 gallons of liquid that would be converted to chlorine dioxide gas.

The long-awaited process was finally under way. The gas was pumped into the building and maintained at high levels for 12 hours in hopes that it would reach into every corner and crack. In the weeks after, 5,029 air and surface samples were taken to determine if any spores survived. In March 2003, Tom Day held a press conference at which he happily announced that the testing found no evidence of live spores. "Not a one," he emphasized. He hoped that dismantling the tubing and chemical tanks could begin soon and that the building would reopen later in the year. How many employees would return remains uncertain. Dena Briscoe, president of Brentwood Exposed, a support group of postal employees who had worked at the Brentwood center, declared: "The majority of workers have anxieties about going back."

A few months earlier, in November 2002, Day had traveled to the Hamilton, New Jersey, municipal hall to appear before the State Assembly's Committee on Homeland Security. He sat in a high-back purple chair facing a semicircle of the six state legislators. As soon as decontamination at Brentwood was complete, he told them, cleanup would begin at the Hamilton center. Assemblyman Gary Guear, vice chairman of the committee, asked Day: "Can you guarantee that the New Jersey facility will be as clean as the Hart Office Building?"

Day answered, "We'll aim for zero contamination." "How hot is the New Jersey facility compared to Brentwood?" Guear asked. "Pretty much the same. They have spores throughout," Day responded.

Assemblywoman Joan Quigley, chairwoman of the committee, asked about chlorine dioxide. How dangerous is it to people? she wanted to know. Day tried to be reassuring: "Even in the worst case, if there were massive leakage of the gas, the advice is just to go inside somewhere and seek shelter, because it dissipates in a half hour or so." As the meeting adjourned the mood was somber. The legislators thanked Day but were still uneasy about the absence of a timetable for cleaning up the Hamilton center.

Meanwhile, the American Media building in Boca Raton, Florida, like the Hamilton facility, remained contaminated and sealed. In February 2003, Congress voted to purchase the structure for $1.00. Estimates to decontaminate the building ranged from $10 million to $100 million, but no one knew when the effort would be undertaken.

BIOPREPAREDNESS

Dr. Shmuel Shapira, deputy director general of Hadassah Hospital in Jerusalem, seemed relaxed despite the anticipation of an American-led invasion of Iraq. It was late February 2003, and he remembered well that during the first Gulf War, 12 years earlier, Iraq had fired 39 Scud missiles into Israel. Initial fears that the Scuds might be carrying biological or chemical poisons turned out to be unfounded. Now, however, worries had revived. Citizens were lining up around the country to obtain gas masks and other paraphernalia in anticipation of a germ or gas attack.

Shapira is in charge of mass casualty coordination for the hospital. He mentioned Israel's particular expertise in this area. "Unfortunately, we've had a lot of experience these past 2 years," he said. In the previous 25 months, 2,100 victims of terrorism were brought to Hadassah Hospital, one of about 30 major hospitals in Israel. None of the attacks involved nonconventional weapons, but they provided valuable, if bitter, training experience. "We handled as many as 100 terror victims in a single day at our hospital alone," he said.

Shapira mentioned the array of preparations in place for a surge of victims of chemical or biological weapons—from the line of 50 outdoor showers to wash them down to stocks of antibiotics and antidotes. In minutes, lounges, foyers, and corridors could be converted into emergency facilities. Extra beds and medications were at the ready, stored below ground level in the hospital's buildings. Embedded in the ceilings of some lobbies was electrical wiring along with pipes that carried oxygen, water, and other necessities for emergency care. Rows of tubes could be dropped in an instant from the ceiling next to newly placed beds. Dr. Shapira said that the 1,050-bed capacity in Hadassah's complex of buildings could be doubled in an hour.

Although trained as an anesthesiologist, Shapira is now engaged largely in hospital management. His tan wooden desk seemed outsized for his small office. As he hung his arm over the back of his chair, his tie flopped below his button-down collar. Tufts of black hair above his ears framed a bald center. In accented English,

Shapira answered my questions: Is Israel doing anything differently in anticipation of an Iraqi attack than it did 12 years ago?

"We were quite prepared for a chemical attack then but not really for biological. We're much better prepared for a bioattack now."

How so?

"In the first place, we have more stocks of antibiotics to treat agents like anthrax and plague. But also, we're ready for smallpox, which was not the case in 1991."

At the time we spoke, 18,000 Israeli first responders had been vaccinated for smallpox with plans to vaccinate another 22,000. Several hundred vaccinated personnel were having their blood drawn to extract something called Vaccinia Immune Globulin, or VIG. This is a substance in the blood of vaccinated people that can be used to treat individuals who suffer adverse reactions from the vaccine's live *vaccinia* virus.

Stockpiles of smallpox vaccine were in place around the country. If a single case of smallpox were diagnosed in Israel, schools, recreation halls, and gymnasiums would immediately be turned into vaccination clinics. Trained crews would work day and night to vaccinate all 6.2 million Israelis in only 4 days. (A vaccination administered within 4 or 5 days of the time of exposure to smallpox still provides protection or reduces the severity of illness.)

Do citizens know where to go for vaccinations? I asked. "No, they don't know now," Shapira responded. But if a case were discovered, "people will get instruction through the media—where to go, when to go." Shapira's voice is muscular, adding a sense of authority to his description. "For instance, if the family name begins with A to H, you go to such-and-such location until 7 p.m." The plan is in place, he said. Then an afterthought: "Certainly people will be scared. There will be anxiety."

Speaking of anxiety, I asked if he was particularly anxious about a biological event. He admitted to worrying, even while maintaining that the threat level was low. He spoke again of Israel's experience with "conventional" terrorism. He also noted that doctors have had occasional experience with insecticide poisoning. (Some insecticides are organophosphates, in the same chemical class as nerve agents, though far less potent.) But biological warfare agents are a different matter: "I mean, there are some agents that no one has probably seen. I am not sure if there is even one physician in Israel who has ever seen smallpox. Anthrax? Cutaneous, maybe, but not inhalation anthrax."

Boaz Ganor, an Israeli terrorism expert, told me that America's bout with bioterrorism in 2001 was followed closely in his country. "Definitely, the anthrax letters raised the awareness of people here."

While in Israel I visit a Patriot missile base on the outskirts of Tel Aviv. Nearly 300 American soldiers had recently arrived to work with Israeli counterparts in anticipation of a missile attack from Iraq. I sat with some of the Americans over lunch in a large white tent. The tent was 200 feet from a communications center—a portable station with computers and launch controls. The station was covered with a huge swath of olive-colored camouflage netting. "I'm Tony Baez from Buffalo, New York," said a strapping young soldier between bites of beef and rice. His smile glistened with large white teeth. Like the other G.I.s, his uniform was a maze of brown and green, pants tucked into high-laced boots.

I asked about the small duffel bag under the table next to his feet. "It's my protective suit," Specialist Baez answered. Similar duffels were carried by every soldier there, ready to be opened on a moment's notice. They rehearse several times a day. "We can put the mask on in 8 seconds," said another soldier at the table. The mask was part of a hood that drapes over the shoulders and is secured to the face underneath by a rubber strap. "Putting on the rest of the gear takes 9 minutes," he added. It takes that long because the outer garment needs to be fastened in orderly sequence with snaps, straps, and ammunition belt.

Outside the tent a Captain Smith pointed to two launch pads in the ravine 50 yards away. The pads were facing east, the direction from which Iraqi missiles would be fired. At what point in the trajectory would the Patriots intercept an incoming Scud? I asked. "We'd hope to make a hit at least a thousand feet above the ground, no lower than 500 feet," he answered. The assumption was that any biological or chemical agents would be destroyed or dissipated by an impact that high. He added a word of assurance that the Patriots in Israel now were more reliable and accurate than those used in the 1991 Gulf War.

The Americans soldiers and their Israeli hosts seemed resolute. There was broad understanding that whether in the form of military assault or terrorism, Israel was quite prepared to deal with a biological or chemical attack. The situation in the United States, with its vastly larger population and territory, was more problematic.

In the spring of 2002 the Johns Hopkins Center for Civilian Biodefense Strategies moved into expanded quarters. On the 8th floor of a red brick office building facing Baltimore's Inner Harbor, the center's staff of 20 churns out position papers and reports. Articles, book reviews, and listings of events and conferences appear in *Biodefense Quarterly*, which the center has published since 1999. The center's leaders were principal contributors to highly influential articles in the *Journal of the American Medical Association* on managing bioterrorism attacks. D. A. Henderson, when director, was the lead author of "Smallpox as a Biological Weapon" and Tom Inglesby, deputy director, of "Anthrax as a Biological Weapon." Tara O'Toole, the current director, was among the coauthors of both.

Soft green carpeting runs throughout the Hopkins center into cubicles and offices that line the corridors. On an overcast day in October 2002, I joined with seven of the professional staff in the conference room for lunch and discussion. Dr. O'Toole said the center is most concerned about developing protection "from an aerosolized mass-casualty attack." Wearing a green cardigan sweater and black shirt, she spoke earnestly. "I want our national leaders to understand that biological weapons are a strategic threat to the United States, and that hasn't happened yet." None of the staff questioned her assessment.

O'Toole's negative view of the country's response to the anthrax attacks in 2001 is unforgiving: "I think the CDC [Centers for Disease Control and Prevention] was terrible. The official response was a national security travesty." Her blond hair was closely trimmed and her smile was taut. Her voice remained at low volume, but the intensity of her delivery filled the room. I ask about the new leadership at CDC, the new director. She responded:

> I am not criticizing any particular individual. I know many of them, and they're trying their best. But they are part of a system that is insular and breeds arrogance. I know Julie Gerberding, and I think she is trying her best in a difficult situation.

O'Toole is distressed that few people appreciate the dimensions of the bioterrorism threat and that the nation is not positioned to respond to an attack.

> We have numerous separate public health departments in the United States. They are mainly unconnected to each other, and most are not prepared for any sort of bioattack.

Monica Schoch-Spana, a senior fellow at the biodefense center, says that after the September 11 attack the center began getting phone calls from people worried about biothreats. "You know, there wasn't even a 'FAQ' [list of frequently asked questions and answers] on bioterrorism on the CDC Web site until after the anthrax outbreak." O'Toole adds that neither the CDC nor any other official agency has produced an after-action report.

Tom Inglesby entered the conversation. "In the first few days after the anthrax attack, the CDC did a credible job. They closed the loop with Larry Bush and Jean Malecki in confirming the first anthrax case in Florida." Inglesby, like O'Toole, is a physician—his medical degree is from Columbia University; hers from George Washington University. He is thin, scholarly; he also holds an appointment in clinical medicine at Johns Hopkins. Wearing a tan shirt and brown tie, he spoke rapidly. "But after the early days, the CDC did not respond well. They didn't get information out quickly, and often they did not get accurate information out at all. That's the concern here."

I asked if everyone at the table agreed. Affirmative nods—everyone was on message: The nation lacks a coordinated program to protect the citizenry. So what exactly should we be doing? I looked to O'Toole. Until then her comments had flowed seamlessly. Now she glanced at the wall beyond the end of the table and reflected. She honed her answers after each of my prompts.

"First, we need to hear an announcement of sustained commitment," O'Toole said.

From whom? I asked.

"From the president. Researchers need to know that funding for their work will continue beyond the short term."

What else?

"A more explicit articulation of the nature of the threat." She related this to her conviction that people do not understand the gravity of the situation. "And then," she continued, "we need the medical and public health communities better prepared. They should be ready for mass casualties."

Is there more?

"We need local as well as national stockpiles of medicines in case of a biological attack. We need to expand the capacity of the public health system."

O'Toole scorned current research approaches to bioterrorism. "We need to marshal our best scientists for this, but that hasn't happened." A lot of money is now supporting work that is not especially imaginative, she said, and for research that is not particularly helpful. The research has been labeled "bioterrorism" so that it could draw funding.

Coming away from the Hopkins biodefense center, one feels appreciation for the valuable studies that its staff has been producing. Further, their concerns about preparedness echo the conclusions of a task force of the Council on Foreign Relations that were released around the same time. (The task force was chaired by former Senators Gary Hart and Warren Rudman, and its report was darkly titled *America Still Unprepared—America Still in Danger.*) The American people remain vulnerable to a bioattack. But the notion that the nation is little better prepared now than during the anthrax attack surely is an exaggeration.

For Kenneth Bernard the difference in preparedness is "like night and day." "We are dramatically better prepared now," he said. Dr. Bernard, 54, is an assistant surgeon general with the U.S. Public Health Service. Since November 2002 he has been director of health and bioterrorism in the White House Homeland Security Council. In March 2003 we met in his office. Above his desk, to the right, hung a poster of Mount Everest, a reminder of his time in the 1980s when he was a health adviser to the Peace Corps in Nepal. Now he is consumed with the weighty challenge of helping to protect this country.

What exactly does his job entail? I asked. "My overall responsibility is to coordinate the White House oversight on health and bioterrorism and biodefense-related issues," he answered.

Dr. Bernard, hair sprinkled with gray, leaned back and rubbed his eyes beneath thinly framed glasses. He puts in long hours but expressed satisfaction with the progress he has seen. He fired off an agenda of issues he attends to, from checking on food safety and vaccine development to securing congressional appropriations for biodefense programs. "Lately I've been working on the BioWatch and BioShield programs," he said, in his crisp manner of speech. Both are recent initiatives announced by President Bush. BioWatch involves developing sensors for distribution around the country to detect a possible biological attack, and BioShield supports development of medical treatments of diseases caused by biological weapons.

Bernard amplified on his assertion that the country is now better prepared:

> If you look at the heightened interest of state and local health departments, the amount of money being spent, the improvement in our surveillance, laboratory networks, research and development for counter-terrorism—in just about every phase it's better.

Bernard's sanguine view was echoed by John Iannarelli, an agent of the Federal Bureau of Investigation. Now in the bureau's public communications office, Iannarelli previously had been a counter-terrorism field agent. He readily acknowledged that coordination among government agencies was weak in the fall of 2001. "But since 9/11, we are in continuous contact with our counterparts in police, fire, and other local agencies," he said. We spoke in the spring of 2003, and he noted that 1,245 FBI agents were now working with 650 state and local police agencies in joint terrorism task forces. The 60 JTTFs in every region of the country were double the number that existed 18 months earlier. "We are dramatically better prepared to deal both with prevention and response," he said emphatically.

Another initiative that could have been mentioned is the new Emergency Command Center in the Office of the Secretary of Health and Human Services. Commander Roberta Lavin has worn the navy blue uniform of the U.S. Public Health Service since 1991. Trained as a nurse practitioner, she had been chief of field operations for immigration and health services. "And then on October 12, 2001, they brought me over here." Her eyes swept across the large work area that had not existed before she arrived. Down the hall from Secretary Thompson's office, Lavin heads the Emergency Command Center, which connects HHS by computer and telephone to local medical response teams around the country.

The Emergency Command Center was ordered set up just after Bob Stevens's anthrax was connected to bioterrorism. A week later the project was under way, and soon after, computer screens were blinking with data. On a late October day in 2002, operators were sitting in front of a half dozen of the 16 computers lined up in rows of four. Lavin offered a smile of assurance and said, "If there was a special reason, an emergency, all the seats would fill quickly."

On the front wall of the Command Center were two 6-foot-wide television screens, one tuned to CNN, the other to Fox News. On the wall to the right, another large screen displayed dozens of names, addresses, and contact telephone numbers. In a Tennessee drawl Lavin explained that the image on that screen was the default image on the computers. But each computer can scroll up or down to show hundreds of other names. "The names are of DMAT leaders and other local team members," she said. Lavin stretched the acronym to sound like "deemat," which stands for "disaster medical assistance team." Beyond these lists, the Command Center operators are in continuous contact with Web sites and emergency health facilities around the world. D. A. Henderson, who had been standing silently behind us, piped in: "They are monitoring for any unusual health incidents, anywhere—an unusual disease, an unusual number of illnesses."

The Command Center provides the HHS secretary with immediate access to 72 DMATs around the country. Each team would be expected to field 35 members in case of emergency. A team has about 100 members, so there is plenty of backup. The computer shows which members at any moment are on alert, which are otherwise available, and which are unavailable. Lavin explained:

> Each DMAT is local, and its members are based in a particular community. They respond to local emergency medical needs. But if there is a disaster they are federalized. They could be moved to a specific location in 6 hours. So they would become part of a federal response just like CDC would. We help to coordinate the response.

Has the center ever had to activate any teams? I asked. "Oh yes," Lavin answered. "We anticipated possible emergencies a few times this past year. The most recent was from a major hurricane. We predeployed several emergency response teams for that."

How about for a biological attack? "Not for any actual biological incident in process because there has not been any since the fall of 2001," she answered. "But we have activated for preplanned events like the IMF." Lavin was referring to recent meetings of the International Monetary Fund in Washington, D.C., at which some protestors were violent. The HHS Emergency Command Center, along with other federal agencies, had response teams in place for possible disruptions, including bioterrorism.

Henderson expressed satisfaction with the command center and spoke of additional activities. He and others in the Office of Public Health Preparedness have been overseeing the distribution of more than $1 billion from HHS to the states for bioterrorism readiness. Under Henderson and Jerry Hauer's direction, the office developed a list of 17 criteria that each state had to meet to receive funding. States began submitting requests in June 2002 after start-up money became available. The criteria range from designating a state director of bioterrorism preparedness to producing time lines for preparedness plans. The aim is to enhance communications between laboratories and officials, drug distribution, hospital readiness, and more.

"I've been pleasantly surprised by how quickly the states have responded," Henderson said. He mentions as well the buildup of the National Pharmaceutical Stockpiles. Twelve repositories around the country each contain 96 tons of materials—antibiotics, vaccines, surgical supplies—available in case of a mass casualty attack. "It takes a Boeing 747 to move one complete stockpile," Henderson said. The materials could be brought to a location in 6 to 12 hours.

All this is not to ignore gaps in the nation's ability to respond to a bioattack. Training and equipping police, fire, and emergency medical personnel to deal with bioterrorism must be continuous. While funds have become available to deal with a range of biothreat issues, some people worry that the burst of support may not be longstanding. "We're doing better than we were," Henderson said, "but preparedness is a long-term process and I'm not sure everyone realizes that."

A sustained process is exactly what health officials at local levels are thinking about. In November 2002 the National Association of County and City Health Officials, NACCHO, reported the results of a survey titled "One Year Later: Improvements in Local Public Health Preparedness Since September 11, 2001." It was based on 342 responses from the 1,600-odd local public health agencies around the country. "Nearly all," the report said, "are moving forward to prepare their communities for bioterrorism." And there was common recognition that preparedness was a multiyear task that needed sustained investment.

About half the respondents indicated that during the previous year they had improved their communications with fire, police, and emergency personnel. Several had engaged in drills and had obtained new technology and equipment. But many felt their staff size was inadequate and that they needed more training.

Six month later, in March 2003, I spoke with the directors of several local agencies. The collective message was that preparations to deal with an attack were better than 18 months earlier but still lacking in some areas. Dr. Lloyd Novick, health commissioner for Onondaga County in upstate New York, said his office had improved disease surveillance and developed a good community response plan for the half million people he serves. "But whether the preparedness is satisfactory is still a question," he said.

For Brian Letourneau, health commissioner for Durham County, North Carolina, "There is no question that we're better off. We routinely meet with police, fire, hospitals, all the players." Still, he worries about the chaos that might ensue if there were a mass evacuation. Carol Moehrle, public health director for five northern Idaho counties, received $400,000 to enhance biopreparedness for the 120,000 people in her district. Her department has been able to add five positions, including an epidemiologist and information and planning personnel. "We are doing very well," she said.

Not all officials were as sanguine. Ron Osterholm is public health director in Cerro Gordo County, Iowa (population 46,000). He echoed the same concerns as Dr. Rex Archer, who heads the health department of Kansas City, Missouri, and is responsible for a population of 500,000. Both said that awareness and understanding in their communities about bioterrorism are much increased. But resources remain scarce. Archer noted the irony that his department added six people to work on bioterrorism through new federal funding. But in the past 18 months, because of budget cuts, 30 other positions were eliminated, including five in the sexually-transmitted diseases area. "Every few months we've been losing some of our disease detectives," he said.

From her vantage point as the bioterrorism coordinator for NACCHO, Zarnaaz Rauf suggested that the mixed impressions I received were quite representative. "Most local officials do feel better prepared than a year and a half ago," she said, "but they're not where they want to be." Still, as Anthony Fauci, director of the National Institute of Allergy and Infectious Diseases, told me, the anthrax letters "triggered a constellation of events that have moved us forward dramatically."

REASSERTING THE NORM

Tibor Tóth, a Hungarian diplomat, is a tall man who takes lengthy strides. His warmth and energy are more evident in private conversation than during the long, tedious sessions over which he has been presiding since 1994. That was the year that several nations began meeting under United Nations auspices to try to strengthen the 1972 Biological Weapons Convention. About 60 of the 147 state parties to the treaty have been participating. For 6 years the Ad Hoc Group, as it is called, had been gathering three or four times a year for week-long meetings in Geneva.

The 1972 convention was the first international agreement to ban an entire category of weapons. Established 3 years after the United States renounced its own offensive biological weapons program, the agreement requires all parties to the treaty to renounce theirs. Those states undertake "never in any circumstances to develop, produce, stockpile or otherwise acquire or retain" biological agents or toxins—except in quantities and types for peaceful purposes. The noble aim of the treaty is enshrined in its determination "for the sake of all mankind, to exclude completely the possibility of bacteriological (biological) agents and toxins being used as weapons."

A broadly recognized weakness of the treaty is its silence on how to verify compliance. Unlike the more recent Chemical Weapons Convention, which was established in 1993 to ban chemical weapons, the biological treaty contains no formal provisions for monitoring or independent inspections. It simply urges that nations consult and exchange information with one another and that allegations of noncompliance be brought to the attention of the UN Security Council. Since 1975, when the biological treaty went into effect, conferences of member states have been held every 5 years to review its effectiveness.

Successive review conferences determined that the treaty was hobbled by the absence of provisions to promote compliance. By the mid-1990s, these concerns intensified as information surfaced about the huge scope of the former Soviet and Iraqi biological programs. Those two countries had been violating the convention even though the Soviet Union was one of the original parties to it and Iraq had signed it. U.S. intelligence estimated that they were among at least 10 countries that were developing germ weapons. These revelations helped give rise to the Ad Hoc Group. After the Fourth Review Conference, in 1996 the Ad Hoc group vowed to produce a protocol on verification within 5 years. That way a document could be considered for adoption at the next review conference, which was scheduled for the end of 2001.

By the time of the final Ad Hoc Group meeting in Geneva,

Switzerland, in July 2001, a 210-page document was in hand. Known as the chairman's text, it was a product of Tibor Tóth's efforts to draw from the mix of proposals and objections offered during the previous 6 years. Many nations anticipated that some version of the text would be approved and sent on to the Fifth Review Conference. Among its key provisions were procedures by which a country would host international inspections of its military and commercial facilities to assure others that they were in compliance with the treaty.

At the July 2001 meeting, more than 50 states indicated general support for the chairman's text. But the United States resisted the tide. Ambassador Donald Mahley, the U.S. representative, declared that the "current approach" was not capable of "strengthening confidence in compliance with the Biological Weapons Convention." The United States thought that cheating could still go undetected, that U.S. biodefense efforts could be jeopardized, and that the confidentiality of proprietary business information might not be protected. When the U.S. delegation indicated it would no longer participate, the meeting collapsed. The United States was strongly criticized, less for failing to support the chairman's text than for its refusal to consider a revised document in *any* form. "We will therefore be unable to support the current text, even with changes," Mahley said.

By the time the Fifth Review Conference convened in November 2001, the United States was reeling from the recent jetliner and anthrax letter attacks. But acrimony about the U.S. position on the protocols remained at high pitch. Before the 3-week conference ended, the divisiveness prompted its suspension without agreement on a final document. A year later, in November 2002, the review conference reconvened. Tibor Tóth, now chairman of the conference, was able to cobble together a consensus for further talks but not about the protocol. With U.S. concurrence, meetings in the next few years would deal with narrowly specified issues, such as ways to mitigate the effects of a bioattack and a code of conduct for scientists.

The U.S. rejection of the protocol continued to prompt criticism. In the January/February 2003 issue of the *Bulletin of the Atomic Scientists*, two biological arms control experts, Mark Wheelis and Malcolm Dando, wrote an article titled "Back to Bioweapons?" The authors speculated that the U.S. rejection may have been inspired by some "offensively oriented" operations that the country wanted kept secret. They based their suspicions on recent disclosures of U.S. activities, such as the production of dried, weaponized anthrax spores for defensive testing.

Alan Zelicoff is a research scientist at the Sandia National Laboratories where he works on the issue of biological weapons nonproliferation. He dismisses the Wheelis/Dando conjectures as baseless. His access to classified documents about the extent of the U.S. biodefense work prompted him to say in an e-mail: "Never (that is to say, not once, not ever) have I had any suspicion that the US biodefense program was intended in ANY way to develop weapons for use on the battlefield."

Indeed, Wheelis and Dando's provocative speculation falls far short of proof that the U.S. is engaged in illegal research. Still, refusal by the United States to negotiate a protocol to strengthen the Biological Weapons Convention creates a risk in its own right. In the absence of an effective international agreement, the proliferation of germ weapons becomes more likely. A strengthened treaty is not just a recitation of rules and procedures. It is a statement of values. It reaffirms the precious international norm that germ weapons cannot be tolerated in civilized society. While sometimes violated in practice, this norm has long historical roots.

Although the specter of biological weapons has grown in recent years, the fact remains that germs have rarely been used as weapons of war or terrorism. In the 20th century the only confirmed use of biological agents against humans in battle was by Japan against China. In the 1930s and 1940s the Japanese dropped ceramic bombs containing plague-infested fleas over Chinese villages from low-flying airplanes. Thousands of Chinese reportedly died from the plague. Apart from warfare, a few instances of bioterrorism have been recorded. But the only known large-casualty event in the United States was the 1984 poisonings with *Salmonella typhimurium* in Oregon restaurants by the Rajneesh cult.

Explanations for the infrequent use of biological weapons range from presumed difficulty in making them to uncertainty about their effectiveness. But another reason deserves to be underscored, the sense of repugnance that these weapons engender.

In pretechnical and advanced societies alike, health has been deemed a self-evident value. In the 4th century B.C., Hippocrates observed that "health is the people's most valuable possession."

More recently, in his landmark book, *On Aggression*, the Nobel Prize-winning Austrian ethologist Konrad Lorenz noted that the "sanctity of the Red Cross is about the only one of the laws of nations that has always been more or less respected by all nations." In all societies, much capital is spent to ward off illness. Thus, efforts to deliberately make people sick contravene a deep-seated human value. No wonder the use of biological weapons has been disparaged as public health in reverse and is deemed abhorrent.

In appealing to the conscience of mankind, the treaties that ban biological and chemical weapons reinforce preexisting inclinations against their use. Strong agreements that provide for verification and punishment of cheaters are as necessary as laws that proscribe other immoral behavior. But salutary policies need not be limited to strengthening and enforcing treaties. Another approach, especially relevant to biological weapons, emerged in the 1980s in the form of a quest for health. It harks back to the smallpox eradication campaign led by D. A. Henderson in the 1960s and 1970s.

In October 1984, in the midst of civil war, President Napoleon Duarte of El Salvador invited Jim Grant for lunch. Grant was the craggy director of the United Nations Children's Fund, UNICEF. He had been presiding over an immunization campaign against polio and other diseases in Colombia, and Duarte wanted him to do the same for children in El Salvador. Journalist Varindra Tarzie Vittachi reported the following conversation in his book *Between the Guns*. With his Cheshire cat grin, Grant said that a cease-fire in the war against the rebels might enable him to implement such a program for Salvadoran children.

Duarte turned to a general next to him: "How long do you think I would remain as president if I asked for a cease-fire as Mr. Grant suggests?" "Oh, about 3 days," the general chuckled. Duarte explained to Grant that calling for a cease-fire would "give too much status to the rebels."

"What if we arrange for the rebels to agree unilaterally not to shoot on those days?" Grant responded.

Duarte contemplated the idea. Instead of a cease-fire, perhaps his troops could observe a period of *tranquillidad*. Thus was born a program called "days of tranquillity."

From 1985 to 1990, when the civil war ended, UNICEF, joined by the Pan American Health Organization and the local Catholic Church, arranged for three cease-fires a year between the government and the guerillas. During these days of tranquillity, the aim was to give every child in the country immunizations and booster shots.

The Salvadoran experience emboldened international health officials to encourage similar truces elsewhere—in Uganda, Lebanon, the Philippines, and Sri Lanka. Whether the immunization ceasefires helped resolve the conflicts is uncertain. But the programs contributed to the worldwide effort to eradicate polio, which declined by 80 percent between 1988 and 1997.

The connection of days of tranquillity to the prevention of biological warfare should be obvious. One can scarcely imagine a program more likely to reinforce the notion of abhorrence about using biological agents for hostile purposes. A party that suspends fighting in order to eradicate disease one day is far less likely to try to spread disease the next.

Days of tranquillity should be celebrated as humanitarian battles *against* pathogens. Of course, heralding such events cannot guarantee good behavior everywhere. But it would graphically reinforce the notion that the use of germs as weapons contravenes an essential value of humanity. Days of tranquillity underscore the notion that biological weapons are unacceptable among civilized people.

The increased threat of bioterrorism, as exemplified by the anthrax letters, is a dismal reminder that that norm is fragile. Strengthening the Biological Weapons Convention would be one way to reaffirm the norm. Programs like the days of tranquillity are another. Severe punishment of offenders is yet another. Some terrorists might try to use germs as weapons anyway. But if institutional barriers are set high and condemnation is assured, many would be less likely to do so.

Meanwhile, the consequences of the 2001 anthrax attacks must still be addressed. Through the pain of the events, much has been learned, though much remains uncertain. Was Ernie Blanco's full recovery—unique among the 11 inhalation cases—attributable to his having received extraordinary doses of ciprofloxacin? Were the two "outlier" cases—the women in New York City and Connecticut who died—unusually susceptible to trace amounts of spores? Will an anthrax-contaminated letter tucked in a kitchen drawer infect someone years from now? Were the 11 inhalation cases identified by the CDC the only such cases?

THE TWELFTH INHALATION CASE?

Surrounded by books and journals, Dr. Tyler Cymet sat in his small office at Sinai Hospital in North Baltimore. He pulled a green sheet of paper from his desk drawer, rolled his chair toward me, and displayed a handwritten list. "These are the symptoms that most of them have complained of," he said, and then read aloud: "memory loss, weakness, weight loss, diaphoresis [perspiration], muscle aches, lymph node swelling, fatigue, depression, increased and decreased temperatures." The list was based on his conversations with survivors who contracted inhalation anthrax in the fall of 2001. As we talked, in March 2003, five of the six survivors were in their second year of illness. "I've examined Leroy Richmond and Qieth McQue," Cymet said, "and I've spoken by phone with David Hose, Norma Wallace, and Jyotsna Patel." He had not contacted Ernesto Blanco, who, at that time, was the only survivor back at work. But he called the others every 3 months to ask, "Is this better? Is this worse? What's going on?"

Tyler Cymet, 40, is an osteopathic physician who completed a residency in internal medicine at Yale University Medical School. Head of family medicine at Sinai Hospital, he is also a professor of internal medicine at Johns Hopkins, 15 minutes from Sinai. He is a marathon runner, and his slender face is framed by a dark beard and mustache. When the telephone rang, he answered softly, "Hi, Tyler here." His first name informality extends to patients and fellow physicians alike. On the wall opposite his desk is a bookcase full of titles like *Pharmacology* and *Principles of Ambulatory Medicine*. He reached to a lower shelf and pulled out a loose-leaf volume bulging with articles and reports about anthrax. Dr. Cymet smiled, "I have three more loose leafs like this one."

What prompted his extraordinary and continuing interest in anthrax? William Paliscak arrived at the emergency room of Sinai Hospital on October 25, 2001. The 37-year-old postal inspector had developed a hacking cough along with heavy perspiration, chills, and severe chest pain. Six days earlier, on October 19, he had been in the Brentwood postal sorting center, before the extent of anthrax contamination there had been recognized. Still, he was given a 10-day supply of Cipro as a precaution. Even as he was taking the pills, Bill Paliscak began to feel ill.

When Dr. Cymet was called to the emergency room, he was surprised to see Allison Paliscak sitting there beside her husband. Allison is the director of health information at Sinai Hospital, where she has worked since 1985. "Hi. What's up?" Cymet asked. Allison introduced him to her husband, who was clearly distressed. "Doc, I can't do anything. I can't breathe. I can't walk. I can't move."

"When did all this start?" Cymet asked.

Paliscak told Cymet that he had begun to feel ill a couple of days earlier, after he had been in the Brentwood postal center in Washington. "Some people there got pretty sick, you know."

By then, Dr. Cymet like people everywhere, knew that four Brentwood workers had contracted anthrax and that two had died.

"I'm scared," Bill Paliscak said. Cymet put a stethoscope to his chest, felt for swollen lymph nodes, and ordered a chest X ray.

When the X ray suggested a possibly widened mediastinum, Cymet immediately had Paliscak admitted to the hospital. Cymet and his associate, Dr. Gary Kerkvliet, suspecting anthrax, ordered more tests and then placed Paliscak on intravenous Cipro. But by week's end the tests for anthrax had come back negative, Bill seemed improved, and he was sent home. A month later his condition deteriorated, and he was in the hospital again.

Previously Bill Paliscak had been in superb physical shape. An athlete who worked out regularly, "he was an avid hockey 'nut,'" his wife Allison said. "He played pick-up games, played in a men's league, wherever he could." That all changed after he went to Brentwood, 4 days following the discovery of the anthrax letter to Senator Daschle. While there, Paliscak, who as a postal inspector is a law enforcement officer, was evaluating the machine that had processed the letter. Cymet and Kerkvliet were lead authors in an article about his case. Appearing in the *Journal of the American Osteopathic Association* (January 2002), the article was titled "Symptoms Associated with Anthrax Exposure: Suspected 'Aborted' Anthrax." The authors noted that:

> as part of the investigation (and while wearing only a simple, store-bought face mask), the patient removed air filters in the area of contamination for evidentiary purposes. In the process of removing and changing these filters, the patient inhaled large quantities of dust particles.

After reviewing his subsequent medical condition, the authors concluded:

> While the patient never met the criteria of positive blood cultures for the diagnosis of anthrax, it is our belief that, despite being culture-negative, the patient manifested definite physiologic changes that do not have any other valid explanation. . . . We strongly believe that

there is a relation between the patient's exposure to anthrax and the symptoms displayed.

Eighteen months after Cymet first saw Bill Paliscak, Cymet said, "He's still sick. He's in the hospital, again, right now." In addition to his other symptoms, Paliscak's arms and legs recently, and unaccountably, had become swollen. Later, in a weak and hesitant voice, Paliscak told me that he cannot get around now without using a walker. "And I can't walk more than 80 feet without getting tired."

Dr. Cymet waved his hand through the air as if tracing the path of a roller coaster. "He's had a rocky course," the doctor said. "I'm trying to figure out where our patient is going and what we should be expecting." Cymet summarizes for me a year and a half of Paliscak's numerous hospital admissions and discharges. After the first 6 months of illness, Paliscak seemed somewhat better. He went home again, but by the summer of 2002 he was back in the hospital. "He had decreased energy, and his blood pressure had fallen to 80 over 40," Cymet said. Normal pressure is 120 over 80. Then in recent months Paliscak began to show some deterioration in brain activity. "That is the same as we've seen in three of the other anthrax patients—low blood pressure and brain involvement," Cymet noted.

I asked if Paliscak had other symptoms similar to those in the other anthrax victims. "Actually he's been *leading*. He's had the symptoms *before* the other inhalation cases," Cymet responded. Why? I wonder. "That might be because he had an incredibly large exposure. He opened the filter of the machine that the mail was being sorted through and spilled the dust on his face."

Dr. Gary Kerkvliet joined our conversation. Unlike the more reserved Dr. Cymet, Kerkvliet is bouncy and ebullient. He laughs easily. Beneath his helmet-style blond hair, his large glasses magnify his light blue eyes. He said he felt frustrated by the CDC's reluctance to say this case is related to anthrax. He rolled his eyes and glanced at Dr. Cymet: "What do you think, Ty?" Cymet offered a silent, knowing smile.

"The CDC kept saying, 'He doesn't have anthrax,'" Kerkvliet said, shaking his head in frustration. "Well, fine," he continued, "but the patient was adamant about wanting to be seen by the CDC." A CDC physician did come up to examine him in December 2001 but never returned.

Since October 2001, Allison Paliscak said, she and her husband have "been living a nightmare." She was grateful for the care and attention Bill had been receiving from Dr. Cymet and Dr.

Kerkvliet. But she was less warmly disposed to the CDC. "After visiting my husband that one time, they seem basically to have dismissed him," she said with a tinge of resentment. For Allison and Bill, the CDC's refusal to allow that his illness is related to anthrax is a signal that the agency is uncaring.

Bill Paliscak's tests were negative for anthrax, and the CDC is surely justified in not calling his case "anthrax." But refusing to classify him as a "suspect" case seems arguable. On October 19, 2001, two weeks after Bob Stevens died, the centers issued an expanded case definition of anthrax. A case would now be deemed "confirmed" if the bacteria could be cultured from a patient's specimen or if two other lab tests were positive for anthrax. The new criteria also allowed for a "suspect" case to be considered even in the absence of a positive test result. That part of the bulky definition reads:

> CDC defines a suspect case as 1) a clinically compatible case of illness without the isolation of *B. anthracis* and no alternative diagnosis, but with laboratory evidence of *B. anthracis* by one supportive laboratory test or 2) a clinically compatible case of anthrax epidemiologically linked to a confirmed environmental exposure, but without corroborative laboratory evidence of *B. anthracis* infection.

Bill Paliscak would seem to be a candidate for the second criterion: clinically compatible signs and symptoms linked to an environmental exposure.

In March 2003, I asked Brad Perkins, one of the CDC's lead investigators during the anthrax outbreak, what the agency's position had been about Mr. Paliscak. "We could never find any evidence that he had anthrax," Perkins said. "We certainly have had an open mind, but we've extensively reviewed his medical records and could not find reason to apply the definition."

As we discussed the question further, Perkins made clear that Paliscak's long-term symptoms had not affected the CDC's thinking. Indeed, the agency does not consider them relevant to anthrax in any of the cases. "We've extensively evaluated all of the survivors that are willing to work with us and found no evidence of long-term sequellae of anthrax."

Perkins acknowledged that "some of the survivors are functioning at lower levels" than a comparable segment of the general population would be. But their disabilities are not "identifiably related to lab tests or clinical findings associated with anthrax." A possible explanation for their current condition, he thought, could be psychological, perhaps because of the trauma of being involved in a bioterrorism incident.

Perkins said that he is, of course, aware of the clause in the case definition that could potentially apply to Paliscak as a suspect case. But he added without amplification, "we have not applied that clause" in defining suspect cases.

Dr. Mary Wright's interpretation is more expansive. She is the principal investigator of a research project at the National Institutes of Health that includes anthrax victims. Since early 2002 she has been overseeing a protocol on the "natural history of anthrax" at the National Institute of Allergy and Infectious Diseases (one of the institutes of the NIH). The purpose is to assess the disease and its effects over time.

Wright has been following the course of Bill Paliscak's illness, along with those of other confirmed and suspected anthrax cases. Her protocol is based on the CDC's definitions, and she *does* consider Paliscak a suspect case. She emphasized that her decision should not be construed as a criticism of CDC. Nor is it an institutional decision by her agency, NIAID. "It's just 'li'l ole me' as the principal investigator who made some clinical judgments about what to use."

Her inclusion of Paliscak as a suspected case is much appreciated by Allison Paliscak and by Drs. Cymet and Kerkvliet, all of whom spoke warmly of her compassion. When sitting with Dr. Cymet, I asked if Paliscak realized he had been exposed to a large volume of dust. "Yes," Cymet answered, and then said something surprising.

"Bill's clothes tested positive and his car tested positive for anthrax."

I had not recalled any mention in his journal article about positive tests results. Did I miss something?

"No, that's right," Cymet answered. Cymet was never shown anything in writing about the clothes or car, so he was reluctant to refer to them in the article. "But two CDC officials visited me and told me about the positive results." Still, all the diagnostic and laboratory tests on Paliscak himself had been negative. "So the CDC says, 'Based on the definition and our criteria, we cannot call it anthrax.'" Cymet shrugged and turned his palm up as if to ask, "So what can I say?"

In an afterthought Dr. Cymet said that he had a public health student research the deaths among all the workers at Brentwood. I mentioned that Dena Briscoe, president of Brentwood Exposed, told me there was an unusual number of deaths there. Cymet said she was correct. " Normally there are only two deaths a year there. We found eight deaths, in addition to those of Mr. Curseen and Mr. Morris, in the one-year period after the anthrax attacks. Four of them had enlarged hearts. We can't explain it."

The numbers are too small to be statistically significant, he said. But clearly he found them suggestive. How could this be related to anthrax? I asked. It was a question that Cymet has asked himself many times. "Maybe because they had lower-dose exposures and our guy had a much higher dose."

Dr. Cymet thumbed through one of his loose-leaf books and pulled out a few pages. He showed me a declassified Army review of some laboratory-associated deaths at Fort Detrick. The first reference was to William Boyles, who, while working with anthrax in 1951, became fatally ill. The review indicated that no anthrax bacteria could be "cultivated" from Boyles's specimens either while he was sick or from his autopsy. But the autopsy report established a diagnosis of anthrax nonetheless. Cymet pointed to a passage that explained why, and he said: "Here's something that shows the spores could cross the blood-brain barrier and stay hidden." From the Army report: "Microscopically, after exhaustive examinations, a few degenerative, long, gram-positive bacilli resembling *B. anthracis* were found only in the Pacchionian granulations (of the brain)." (In the 18th century, Italian anatomist Antonio Pacchioni, identified these tiny elevations on the outer surface of the brain.)

Dr. Cymet realized that some current tests to identify anthrax, like immunohistochemical staining, did not exist in 1951. Still, he found the report intriguing. Could bacilli have been embedded in the brains of Bill Paliscak and others who were exposed to spores from the anthrax letters?

EPILOGUE

The case of William Paliscak is emblematic of much about anthrax that remains a mystery. In the context of public health, even describing his case as anthrax related has been a matter of dispute. In its medical context, as with most of the inhalation anthrax survivors, the course of his difficult illness has been puzzling. In the larger perspective of bioterrorism, it underscores the agonizing consequences that can result from a germ attack. Since the anthrax letters were mailed, victims are still sick, buildings are still contaminated, and many people remain anxious.

The bioterrorism attacks in the fall of 2001 have become a learning exercise, and many previously held assumptions have been turned upside down. Even the quantity of spores necessary to cause inhalation anthrax has come under question. Some individuals who were exposed to huge numbers of bacilli did not become ill. Others, who apparently inhaled only a few organisms, contracted the disease and died.

Before the attack it was commonly believed that a patient would need immediate antibiotic treatment to survive inhalation anthrax. In fact, none of the six inhalation survivors were treated with antibiotics until after they had been sick for several days. Similarly, pre-attack dogma held that a chest X ray of an anthrax victim would almost always reveal a widened mediastinum, the area between the lungs. But X rays of at least four of the 11 inhalation victims showed no mediastinal enlargement.

Perhaps the biggest surprise produced by the anthrax incidents was the effectiveness of the U.S. mail as a means of delivering disease. Not only did the anthrax letters carry their poisons to the target addresses, they left a trail of bacteria along the way. Before the attack, evidently no one considered the fact that 3-micron spores could leak through 20-micron pores—the diameter of microscopic holes in the envelope paper.

While the letters caused illness and death to relatively few people, they disrupted the lives of millions. The effects were bad enough from just those five or six letters. But they hinted at the devastation that could arise from hundreds of letters or from other means of delivery and exposure, or from organisms that cause contagious diseases like smallpox and plague, or from a strain of bug that is resistant to antibiotics and vaccines.

The anthrax letters moved concerns about bioterrorism from theory to reality. Complacency has since given way to a broad recognition of the country's vulnerability to germ weapons. Since the fall of 2001 the nation has been responding to the threat with additional money and programs. Whether in the area of law enforcement, public health, or scientific research, more resources have become available to protect against a biological attack. Opinions differ about how much more is needed and where the money should be spent, but most experts agree that in 2003 the United States was better prepared for a bioattack than it was in 2001.

At the same time, there is an understanding that more needs to be done. A dedicated terrorist could doubtless find a vulnerable location or pathway to release his poisons; if not through the mail, then in a building ventilation system. If not by way of a crop duster, then in a subway. Of course, no amount of money and preparation can ever assure absolute protection, but if sensibly applied, they can help.

On February 5, 2003, Secretary of State Colin Powell appeared before the U.N. Security Council to speak of the danger posed by Saddam Hussein's regime in Iraq. In particular, he deplored Hussein's presumed arsenal of biological weapons. Dressed in a dark business suit with an American flag emblem in his lapel, Powell held up a glass vial of cream-colored powder. Gingerly grasped between his right thumb and index finger, the vial was the size of his pinky. "Less than a teaspoon of dry anthrax, a little bit, about this amount," he said, "shut down the United States Senate in the fall of 2001." (The powder in the vial was not anthrax, but a harmless simulant.)

That the secretary should highlight his concern about Iraqi germ weapons was not surprising. A year earlier in a brief conversation he remarked to me that he considered biological agents to be the most worrisome of all weapons. His deep-seated concern was consistent with that of President Bush. After the anthrax attacks, Bush referred to biological weapons as potentially the most dangerous of weapons. The remarks by these leaders, as well as President Clinton, embrace a general wisdom that biological weapons are especially loathsome.

As a backdrop to preventing their use, reinforcing long-held attitudes about the immorality of these weapons is important. But more immediate prospects for minimizing the chances of their use lie with science. Laboratory tests to identify a bioagent, experiments to learn how infected mail contaminates other objects, research on diagnostics and treatment, and interpreting the mind-set of people who send both false and real biothreats—all point to the role of medical and biological science in preventing and responding to bioterrorism. Continued advances in these areas will diminish the threat of germs as weapons and, in consequence, lessen interest in their use.

Meanwhile, it is incumbent on decent people to continue to condemn the use of these weapons as a debasement of civilized society everywhere.

BIBLIOGRAPHY

The following items represent only a fraction of the vast literature on bioterrorism. Organized by predominant themes in each chapter, they augment the discussion of the anthrax letters and provide a base of information about bioterrorism in general.

1
DEADLY DIAGNOSIS

On Bob Stevens and the suspicion that he had anthrax:

Bhatt, Sanjay. "'No Evidence of Terrorism' in Isolated Case, U.S. Health Secretary Says," *The Palm Beach Post*, October 5, 2001.

"Bob Stevens: The Man Who Saved America," *The National Enquirer*, October 30, 2001.

Canedy, Dana and Nicholas Wade. "Florida Man Dies of Rare Form of Anthrax," *The New York Times*, October 6, 2001.

"The Nightmare That Came in the Mail," *The National Enquirer*, October 30, 2001.

Palm Beach County, Florida Health Department. Interviews with Maureen Stevens by Jean Malecki, Director, October 3, 2001 and October 6, 2001.

On Larry Bush, Jean Malecki, and the confirmation of the anthrax diagnosis:

Bhatt, Sanjay. "Experts, Officials Flummoxed by Nature of Anthrax Attack," *The Palm Beach Post*, November 5, 2001.

Bush, Larry M. et al. "Index Case of Fatal Inhalational Anthrax Due to Bioterrorism in the United Sates," *The New England Journal of Medicine*, November 29, 2001.
Flynn, Sean. "Whatever Happened to Anthrax?" *Esquire*, March 2003.
Malecki, Jean. "Palm Beach Anthrax," *NACCHO Exchange*, Spring 2002.
Traeger, Marc S. et al. "First Case of Bioterrorism-Related Inhalational Anthrax in the United States, Palm Beach County, Florida, 2001," *Emerging Infectious Diseases*, October 2002.
Wiersma, Steven. "Inhalation Anthrax in Florida," presented at the International Conference on Emerging Infectious Diseases 2002, Centers for Disease Control and Prevention et al., Atlanta, GA, March 24-27, 2002.

2
AMERICAN MEDIA

On diagnosing and treating Ernie Blanco's anthrax:

Berman, Noah, Alice Gregory, and Joan Murawski. "Second American Media Worker Contracts Inhalation Anthrax," *The Palm Beach Post*, October 16, 2001.
Evans, Christine. "The Miracle That Saved Ernie Blanco's Life," *The Palm Beach Post*, December 2, 2001.
Inglesby, Thomas V. et al. "Anthrax as a Biological Weapon: Medical and Public Health Management," *Journal of the American Medical Association*, May 12, 1999.
U.S. House of Representatives. Testimony of Carlos Omenaca, M.D., before the Subcommittee on Oversight and Investigations of the Veterans' Affairs Committee. Washington, DC: Government Printing Office, November 14, 2001.

On American Media and suspicions about the anthrax perpetrators:

"Every American's Worst Nightmare by the People Who Lived It," *The National Enquirer*, October 30, 2001.
"The Face of Evil: What Fuels Madman Osama Bin Laden's Hatred of America and Drives Him to Commit Mass Murder?" *Globe*, October 2, 2001.
Kidwell, David, Manny Garcia, and Larry Lebowitz. "Authorities Trace Anthrax that Killed Florida Man to Iowa Lab," *The Miami Herald*, October 10, 2001.
Mahler, Jonathan. "The National Enquirer: A Tabloid Story," in Charles Melcher and Valerie Virga, Editors, *The National Enquirer: Thirty Years of Unforgettable Images*. New York: Talk Miramax Books/Hyperion, 2001.

3
THE NATION AT RISK

On false anthrax alarms:

Leusner, Donna. "Package with White Powder Sparks Trenton Anthrax Scare," *The Star-Ledger* (Newark, NJ), October 11, 2001.

On anthrax cases in New York City and the Washington, D.C. area:

Altman, Lawrence K. "Doctor in City Reported Anthrax Case Before Florida," *The New York Times*, October 18, 2001.

Dewan, Puneet K. et al. "Inhalational Anthrax Outbreak among Postal Workers, Washington, DC, 2001," *Emerging Infectious Diseases*, October 2002.

Emling, Shelley and Scott Shepard, "NBC Aide Contracts Skin Anthrax," *The Palm Beach Post*, October 13, 2001.

Hsu, Vincent P. et al. "Opening a *Bacillus anthracis*-Containing Envelope, Capitol Hill, Washington, DC: The Public Health Response," *Emerging Infectious Diseases*, October 2002.

Layton, Marcelle. "Cutaneous Anthrax in New York City: Expect the Unexpected," presented at the International Conference on Emerging Infectious Diseases 2002, Centers for Disease Control and Prevention et al., Atlanta, GA, March 24-27, 2002.

Martinez, Barbara. "Anthrax Victims' Fate Varied by Their Hospital and Doctor," *The Wall Street Journal*, November 27, 2001.

New York City Department of Health. "Health Department Announces an Anthrax Case in New York City," press release, October 12, 2001.

New York City Department of Health. "Health Department Announces Second Case of Cutaneous (Skin) Anthrax," press release, October 15, 2001.

New York City Department of Health. "Health Department Announces Third Case of Cutaneous (Skin) Anthrax," press release, October 18, 2001.

New York City Department of Health. "Health Department Confirms Fourth Case of Cutaneous (Skin) Anthrax," press release, October 19, 2001.

New York City Department of Health. "Second Employee at NBC Has Probable Case of Skin Anthrax; Suspicious Illness Was Previously Reported by NYCDOH on October 13," press release, October 25, 2001.

On initial diagnoses of anthrax in New Jersey:

Greene, Carolyn M. et al. "Epidemiologic Investigations of Bioterrorism-Related Anthrax, New Jersey, 2001," *Emerging Infectious Diseases*, October 2002.

Pfaff, Leslie Garisto. "Lessons from Anthrax," *New Jersey Monthly*, May 2002.

Verrengia, Joseph B. "Anthrax Mystery and Effort to Solve it Unfolded, Case by Nerve-Racking Case," *Newsday*, October 27, 2001.

4
ULTIMATE DELIVERY: THE U.S. MAIL

On delivering anthrax as a weapon:

Cole, Leonard A. *Clouds of Secrecy: The Army's Germ Warfare Tests Over Populated Areas*. Lanham, MD: Rowman and Littlefield, 1990.

On the postal system and additional anthrax cases in New Jersey:

Pearson, Michele. "Anthrax in the U.S. Postal System," presented at the International Conference on Emerging Infectious Diseases 2002, Centers for Disease Control and Prevention et al., Atlanta, GA, March 24-27, 2002.

Reefhuis, Jennita. "Letters from Trenton: The Anthrax Investigation at the Source, New Jersey, 2001," presented at the International Conference on Emerging Infectious Diseases 2002, Centers for Disease Control and Prevention et al., Atlanta, GA, March 24-27, 2002

Tan, Christina G. et al., "Surveillance for Anthrax Cases Associated with Contaminated Letters, New Jersey, Delaware, and Pennsylvania, 2001," *Emerging Infectious Diseases*, October 2002.

Taylor, Alan. "Poor Richard, Poor Ben," *The New Republic*, January 13, 2003.

U.S. Postal Service. History of the U.S. Postal Service: http://www.usps.com/history/his1.htm

5
THE OUTLIERS

On the diagnosis and death of Kathy Nguyen:

Finkelstein, Katherine E. "Confusion and Anxiety, and Plenty of Antibiotics," *The New York Times*, October 31, 2001.

Holtz, Timothy H. "Inhalational Anthrax—New York City, October-November 2001," presented at the International Conference on Emerging Infectious Diseases 2002, Centers for Disease Control and Prevention et al., Atlanta, GA, March 24-27, 2002.

Kleinfield, N. R. "Anthrax Investigators Are Hoping Bronx Case Leads Them to Source," *The New York Times*, November 6, 2001.

Mina, Bushra et al., "Fatal Inhalational Anthrax with Unknown Source of Exposure in a 61-Year-Old Woman in New York City," *Journal of the American Medical Association*, February 20, 2002.

New York City Department of Health. "Alert #6: Inhalational Anthrax Case in New York City," October 30, 2001.

Revkin, Andrew C. "Detectives Ask for Help in Tracing Victim's Steps," *The New York Times*, November 7, 2001.

Steinhauer, Jennifer. "Hospital Worker's Illness Suggests Widening Threat; Security Tightens Over U.S.," *The New York Times*, October 31, 2001.

On the diagnosis and death of Ottilie Lundgren:

Barakat, Lydia et al., "Fatal Inhalational Anthrax in a 94-Year-Old Connecticut Woman," *Journal of the American Medical Association*, February 20, 2002.

Centers for Disease Control and Prevention. "Update: Investigation of Bioterrorism-Related Inhalational Anthrax—Connecticut, 2001," *Morbidity and Mortality Weekly Report*, Atlanta, GA, November 30, 2001.

Daly, Matthew and Diane Scarponi. "Woman in Rural Town Tests Positive for Inhalation Anthrax," *The Hartford Courant*, November 21, 2001.

Hadler, James L. "Inhalation Anthrax of Unknown Source," presented at the International Conference on Emerging Infectious Diseases 2002, Centers for Disease Control and Prevention et al., Atlanta, GA, March 24-27, 2002.

Williams, Alcia A. et al., "Bioterrorism-Related Anthrax Surveillance, Connecticut, September-December, 2001," *Emerging Infectious Diseases*, October 2002.

Zielbauer, Paul. "Connecticut Woman, 94, Is Fifth to Die from Inhalation Anthrax," *The New York Times*, November 22, 2001.

6
D.A. HENDERSON, THE CDC, AND THE NEW MIND-SET

On D.A. Henderson:

Henderson, Donald A. "Public Health Preparedness," presented at the AAAS Colloquium on Science and Technology Policy, Washington, DC, April 11-12, 2002.

Liu, Mark. "Best Prepare for the Worst: How D. A. Henderson, M.D., M.P.H., Saved the Planet from Smallpox, and How He Plans to Do It Again," *Rochester Medicine*, Spring/Summer, 2002.

On smallpox and its possible use as a biological weapon:

Centers for Disease Control and Prevention. "Notice to Readers: 25th Anniversary of the Last Case of Naturally Acquired Smallpox," *Morbidity and Mortality Weekly Report*, Atlanta, GA, November 13, 2002.

Fenner, F. et al. *Smallpox and Its Eradication.* Geneva: World Health Organization, 1988.

Halloran, M. Elizabeth et al. "Containing Bioterrorist Smallpox," *Science*, November 15, 2002.

Koopman, Jim. "Controlling Smallpox," *Science*, November 15, 2002.

Orent, Wendy. "Smallpox Vaccines as a First Line of Defense," *Los Angeles Times*, October 20, 2002.

Tucker, Jonathan B. *Scourge: The Once and Future Threat of Smallpox.* New York: Atlantic Monthly Press, 2001.

On anthrax, bioterrorism, and the Centers for Disease Control and Prevention:

Altman, Lawrence K. "C.D.C. Team Tackles Anthrax," *The New York Times*, October 16, 2001.

Altman, Lawrence K. "When Everything Changed at the C.D.C.," *The New York Times*, November 13, 2001.

Altman, Lawrence K. "Disease Control Center Bolsters Terror Response," *The New York Times*, August 26, 2002.

Marshal, Eliot. "AIDS Researcher Named CDC Chief," *Science*, July 12, 2002.

McKenna, M. A. J. "First Line of Defense: CDC Foot Soldiers Hit Nation's Streets," *The Atlanta Constitution*, August 30, 2002.

Rienzi, Greg. "Bioterrorism Puts Center in Headlines," *The Johns Hopkins Gazette*, November 5, 2001.
Smith, Stephen. "CDC Sees Progress in Terror Response," *Boston Globe*, September 30, 2002.
Stolberg, Sheryl Gay and Judith Miller. "Bioterror Role an Uneasy Fit for the C.D.C.," *The New York Times*, November 11, 2001.

7
A SCIENTIST'S RACE TO PROTECTION

On science and the detection of anthrax:

Schuch, Raymond, Daniel Nelson, and Vincent A. Fischetti. "A Bacteriolytic Agent that Detects and Kills *Bacillus anthracis*," *Nature*, August 2002.
Watson, James D. *The Double Helix*. New York: Atheneum, 1968.
Wortman, Marc. "DNA Chips Target Cancer," *Technology Review*, July/August 2001.

On Nancy Connell and bioterrorism:

Jacobs, Eve. "Scientists Fight Bioterrorism in the Lab," *Healthstate* (The University of Medicine and Dentistry of New Jersey), Fall 2001.
MacPherson, Kitta. "Gearing Up for the Bioterrorism Battle: UMDNJ Scientist Looks for Defense Against Germs," *The Star-Ledger* (Newark, NJ), October 7, 2001.
Regaldo, Antonio. "Armchair Sleuths: Tracking Anthrax Without a Badge," *The Wall Street Journal*, October 14, 2002.

On science and bioterrorism:

Agres, Ted. "Bioterrorism Projects Boost US Research Budget," *The Scientist*, March 18, 2002.
Anderson, Curt. "Bush Signs 4.6B Bioterror Bill," *The Washington Post*, June 12, 2002.
Atlas, Ronald. "National Security and the Biological Research Community," *Science*, October 25, 2002.
Coughlin, Kevin. "Security Puts Crimp in Anthrax Research: Scientists Frustrated by Federal Delays," *The Star-Ledger* (Newark, NJ), March 31, 2002.
U.S. House of Representatives. Hearing on "The Threat of Bioterrorism in America: Assessing the Adequacy of the Federal Law Relating to Dangerous Biological Agents," before the Subcommittee on Oversight and Investigations, Committee on Commerce. Washington, D.C.: Government Printing Office, May 20, 1999.
Wilkie, Dana. "Today's World: Research vs. Security: Investigators Find It Increasingly Difficult to Obtain Bacterial Pathogens for Antibioterrorism Studies," *The Scientist*, September 30, 2002.

8
TERROR BY HOAX

On bioterrorism hoaxes, threats, and responses:

"APIC Bioterrorism Task Force, Bioterrorism Readiness Plan: A Template for Healthcare Facilities," report presented at conference on bioterrorism, Association for Professionals in Infection Control and Epidemiology, Baltimore, MD, June 24, 1999.

Avila, Oscar and Christine Vendel. "KC Hit by Scare Over Anthrax," *The Kansas City Star*, February 23, 1999.

Centers for Disease Control and Prevention. "Bioterrorism Alleging Use of Anthrax and Interim Guidelines for Management—United States, 1998," *Morbidity and Mortality Weekly Report*, Atlanta, GA, February 5, 1999.

Cole, Leonard A. "Bioterrorism Threats: Learning from Inappropriate Responses," *Public Health Management and Practice*, July 2000.

Dolnik, Adam and Jason Pate. "2001 WMD Terrorism Chronology," Center for Nonproliferation Studies, Monterey Institute of International Studies, September 12, 2002.

Federal Bureau of Investigation, WMD Operations Unit. "Anthrax Advisory," *www.emergency.com/fbiantrx.htm,* posted December 1998.

Federal Emergency Management Agency. U. S. Fire Administration, *Fire Department Response to Biological Threat at B'nai B'rith Headquarters*, Washington, DC, April 1997.

Kansas City, Missouri, Fire Department. "After Action Report, 1001 E. 47th Street (Planned Parenthood)," interdepartmental communication, March 1, 1999.

Smith, E.N. "Two Packages Purporting to Contain Anthrax Cause Disruptions," *The Atlanta Journal-Constitution*, February 4, 1999.

Wald, Matthew L. "Suspicious Package Prompts 8-Hour Vigil at B'nai B'rith," *The New York Times*, April 25, 1997.

On anthrax hoax perpetrators:

Federal Bureau of Investigation. Statement by Jeffrey J. Berkin, Assistant Special Agent, before the House Committee on Governmental Reform, Subcommittee on Government Efficiency, Financial Management, and Intergovernmental Relations. Washington, DC: Government Printing Office, July 1, 2002.

Jones, Meg. "Anthrax Hoaxes Continue in Milwaukee, Racine," *Milwaukee Journal Sentinel*, January 12, 2000.

Moreno, E. Mark. "Worker in Court Over Anthrax Call," *San Jose Mercury News*, January 1, 1999.

Rosenzweig, David. "Accountant Fined for Anthrax Hoax," *Los Angeles Times*, July 22, 1999.

"Two Teens Convicted of Felonies in Anthrax Hoax," Associated Press, *The Owensboro Messenger-Inquirer* (Kentucky), March 10, 1999.

On Clayton Lee Waagner and hoax terrorism:

"Clayton Lee Waagner Apprehended," *Anti-Abortion Violence Watch*, February 25, 2002.

Horsley, Neal. "Clayton Waagner Explains How and Why He Sent Hundreds of Fake Anthrax Letters to the Abortion Industry in the USA," *www.christiangallery.com/waagneranthrax.htm*, December 2, 2001.

Roddy, Dennis B. "Abortion Clinic Stalker Arrested in Ohio," *Post-Gazette* (Pittsburgh, PA), December 6, 2001.

Slobodzian, Joseph A. "Suspect Indicted for Threatening Abortion Clinics with Anthrax," *Philadelphia Inquirer*, September 19, 2002.

U.S. District Court for the Eastern District of Pennsylvania. U.S.A. v. Clayton Lee Waagner, indictment, September 2002.

9
WHO DID IT?

On the supposition of a domestic perpetrator:

Broad, William J. and Judith Miller. "Anthrax Inquiry Looks at U.S. Labs," *The New York Times*, December 2, 2001.

Couzin, Jennifer. "Unconventional Detective Bears Down on a Killer," *Science*, August 23, 2002.

Ember, Lois. "An Intellectual Provocateur; Chemist Barbara Hatch Rosenberg Has Been Active in Pivotal Societal Issues for Over Two Decades," *Chemical and Engineering News*, April 1, 2002.

Federal Bureau of Investigation. Amerithrax Press Briefing, "Linguistic/Behavioral Analysis of Anthrax Letters," Washington, DC, *www.fbi.gov/anthrax/amerithrax.htm*, posted November 9, 2001.

"Focus of U.S. Anthrax Probe Is Domestic—Ridge," Reuters, January 13, 2002.

Kristof, Nicholas D. "Profile of a Killer," *The New York Times*, January 4, 2002.

Kristof, Nicholas D. "Connecting Deadly Dots," *The New York Times*, May 24, 2002.

Kristof, Nicholas D. "Anthrax? The F.B.I. Yawns," *The New York Times*, July 2, 2002.

Kristof, Nicholas D. "The Anthrax Files," *The New York Times*, July 12, 2002.

Kristof, Nicholas D. "Case of the Missing Anthrax," *The New York Times*, July 19, 2002.

Kristof, Nicholas D. "The Anthrax Files," *The New York Times*, August 13, 2002.

Lemann, Nicholas. "The Anthrax Culprit," *The New Yorker*, March 18, 2002.

Miller, Judith and William J. Broad. "F.B.I. Has a 'Short List' of Names in Its Anthrax Case, the U.S. Says," *The New York Times*, February 26, 2002.

Nass, Meryl. "In Search of the Anthrax Attacker," *www.redflags.com*, posted February 11, 2002.

Rosenberg, Barbara Hatch. Federation of American Scientists, "A Compilation of Evidence and Comments on the Source of the Mailed Anthrax," *www.fas.org*, posted November 12, 2001, revised December 10, 2001.

Rosenberg, Barbara Hatch. Federation of American Scientists, "Analysis of the Source of the Anthrax Attacks," *www.fas.org*, posted January 17, 2002.

Rosenberg, Barbara Hatch. Federation of American Scientists, "Is the FBI Dragging Its Feet?" *www.fas.org*, posted February 5, 2002.

Shane, Scott. "Bioterrorism Riddle, $1.25 Million Reward Stimulates Interest," *The Baltimore Sun*, January 6, 2002.

On Steven Hatfill:

Altimari, Dave, Jack Dolan, and David Lightman. "The Case of Dr. Hatfill: Suspect or Pawn?" *The Hartford Courant*, June 27, 2002.

Dishneau, David. "Anthrax Probe Back Near Hatfill's Ex-Home," Associated Press, January 27, 2003.

Rozen, Laura. "Who Is Steven Hatfill?" *The American Prospect*, June 27, 2002.

Schemo, Diana Jean. "Weapons Expert Attacks F.B.I. and Ashcroft on Anthrax Inquiry," *The New York Times*, August 26, 2002.

On the supposition of a foreign connection:

Bhatt, Sanjay, Eliot Kleinberg, and Alice Gregory. "Anthrax, Al-Qaeda Linked by Leg Sore?" *The Palm Beach Post*, March 24, 2002.

Fainaru, Steve. "Officials Continue to Doubt Hijackers' Link to Anthrax; Fla. Doctor Says He Treated One for Skin Form of Disease," *The Washington Post*, March 24, 2002.

Gordon, Michael R. "U.S. Says It Found Qaeda Lab Being Built to Produce Anthrax," *The New York Times*, March 23, 2002.

Tell, David. "Remember Anthrax? Despite the Evidence, the FBI Won't Let Go of Its 'Lone American' Theory," *The Weekly Standard*, April 29, 2002.

U.S. House of Representatives. Testimony of Richard O. Spertzel, "Russia, Iraq, and Other Potential Sources of Anthrax, Smallpox, and Other Bioterrorist Weapons," before the House Committee on International Relations. Washington, DC: Government Printing Office, December 5, 2001.

U.S. House of Representatives. Testimony of Richard O. Spertzel, "State of the Iraqi Weapons of Mass Destruction Program," before the House Armed Services Committee. Washington, DC: Government Printing Office, September 10, 2002.

Woolsey, R. James. "The Iraq Connection," *The Wall Street Journal*, October 18, 2001.

10
LOOSE ENDS

CONTAMINATION

Belluck, Pam. "Anthrax Outbreak of '57 Felled a Mill but Yielded Answers," *The New York Times*, October 27, 2001.

Bushouse, Kathy. "Congress Agrees to Buy, Clean Up Anthrax-Tainted AMI Building," *South Florida Sun-Sentinel*, February 14, 2003.

Covert, Norman M. *Cutting Edge: A History of Fort Detrick, Maryland, 1943-1993*, U.S. Army Garrison, Fort Detrick, MD, 1993.

Fernandez, Manny. "A Patient Assault on Anthrax," *The Washington Post*, December 18, 2002.

Fernandez, Manny. "No Anthrax Spores Found at Brentwood," *The Washington Post*, March 5, 2003.

Harrison, David. "Legacy of Fear on Blighted Anthrax Island," *The Daily Telegraph*, October 14, 2001.

Lipman, Larry. "Utah Congressman Fights Anthrax Building Deal," *The Palm Beach Post*, February 11, 2003.

Pearson, Graham S. "Gruinard Island Returns to Civil Use," *The ASA Newslettter*, no. 86, Applied Science and Analysis, Inc., Portland, ME, 2001.

"Reopening of Senate Building Is Delayed," *The New York Times*, January 18, 2002.

Schmid, Randolph E. "Anthrax Cleanup Set at D.C. Postal Facility: Price Tag Put at $22 Million," *The Record* (Hackensack, NJ), June 16, 2002.

Shane, Scott. "Anthrax Fighters Await Outcome. Brentwood: After Months of Preparation, the $100 Million Fumigation of the D.C. Mail-Sorting Center is Done. Will Tests Show the Toxic Spores Have All Been Killed?" *The Baltimore Sun*, December 27, 2002.

Winston, Sherie. "D.C. Anthrax Cleanup Cost Jumps 80%," *Industry Headlines* (McGraw-Hill Construction), October 17, 2002.

BIOPREPAREDNESS

Altman, Lawrence K. and Gina Kolata. "Anthrax Missteps Offer Guide to Fight Next Bioterror Battle," *The New York Times*, January 6, 2002.

Davidson, Keay. "U.S. Gears Up for Smallpox Threat: Nation Is Much Better Prepared Than It Was Last Year, Official Says," *San Francisco Chronicle*, December 6, 2002.

Eban, Katherine. "Waiting for Bioterror," *The Nation*, December 9, 2002.

Hart, Gary and Warren B. Rudman, Co-chairs, "America Still Unprepared—America Still in Danger," report of an Independent Task Force, sponsored by the Council on Foreign Relations, New York, NY, 2002.

Meckler, Laura. "Nation Makes Major Strides, Yet Much Work Remains, in Preparing for Bioterrorism," Associated Press, September 18, 2002.

Miller, Judith. "U.S. Deploying Monitor System for Germ Peril," *The New York Times*, January 22, 2003.

Shapira, Shmuel C. and Joshua Shemer. "Medical Management of Terrorist Attacks," *Israel Medical Association Journal*, July 2002.

REASSERTING THE NORM

Chevrier, Marie Isabelle. "Waiting for Godot or Saving the Show? The BWC Review Conference Reaches Modest Agreement," *Disarmament Diplomacy*, December 2002-January 2003.

Cole, Leonard A. "The Poison Weapons Taboo: Biology, Culture, and Policy," *Politics and the Life Sciences*, September 1998.

Hay, Robin. "Humanitarian Ceasefires: An Examination of Their Potential Contribution to the Resolution of Conflict," Working Paper 28, Institute for Peace and Security, Ottawa, Ontario, July 1990.

Higgins, Alexander G. "Group Suspends Negotiations After U.S. Pullout," Associated Press, *North County Times* (San Diego County, CA), August 4, 2001.

Vittachi, Varindra Tarzie. *Between the Guns: Children as a Zone of Peace.* London: Hodden and Stoughton, 1993.
Walker, John R. *Orphans of the Storm: Peacebuilding for Children of War.* Toronto: Between the Lines, 1993.
Wheelis, Marc and Malcolm Dando, "Back to Bioweapons?," *Bulletin of the Atomic Scientists*, January/February 2003.
United Nations. Highlights of press conference by Tibor Tóth, Chairman of the Ad Hoc Group of States Parties to the Convention on Biological Weapons, Palais Des Nations, Geneva, July 25, 2001.

THE TWELFTH INHALATION CASE?

Broad, William J. and Denise Grady. "Science Slow to Ponder the Ills that Linger in Anthrax Victims," *The New York Times*, September 16, 2002.
Cymet, Tyler C. et al. "Symptoms Associated with Anthrax Exposure: Suspected 'Aborted' Anthrax," *Journal of the American Osteopathic Association*, January 2002.
Fernandez, Manny and Phuong Ly, "Brentwood Deaths Put Employees on Edge," *The Washington Post*, June 15, 2002.
Report on Demilitarization of Fort Detrick (1977), Appendix F, Fact Sheet on Three Laboratory Occupational Deaths, 1951, 1958, 1959.
Shane, Scott. "Postal Employee Might Have Anthrax, Man Likely Exposed but CDC Tests Find No Anthrax in Blood," *The Baltimore Sun*, January 9, 2002.
Shane, Scott. "Postal Inspector's Severe Illness Defies Diagnosis After 9 Months: He Was Exposed to Spores on the Job, but Tests Fail to Show He Has Anthrax," *The Baltimore Sun*, July 15, 2002.
Stolberg, Sheryl Gay. "Ill Postal Worker Has Symptoms that Stop Short of Anthrax," *The New York Times*, January 11, 2002.

ADDITIONAL BOOKS

Alibek, Ken. *Biohazard: The Chilling Story of the Largest Covert Biological Weapons Program in the World.* New York: Random House, 1999.
Barnaby, Wendy. *The Plague Makers: The Secret World of Biological Warfare.* London: Vision Paperbacks, 1999.
Butler, Richard. *The Greatest Threat: Iraq, Weapons of Mass Destruction, and the Crisis of Global Security.* New York: Public Affairs, 2000.
Cole, Leonard A. *The Eleventh Plague: The Politics of Biological and Chemical Warfare.* New York: W. H. Freeman, 1997.
Cordesman, Anthony H. *Weapons of Mass Destruction in the Middle East.* London: Brassey's, 1991.
Dando, Malcolm. *Biological Warfare in the 21st Century.* London: Brassey's, 1994.
Drell, Sidney D., Abraham D. Sofaer, and George D. Wilson, Editors. *The New Terror: Facing the Threat of Biological and Chemical Weapons.* Stanford, CA: Hoover Institution Press, 1999.
Falkenrath, Richard A., Robert D. Newman, and Bradley A. Thayer. *America's Achilles' Heel: Nuclear, Biological, and Chemical Terrorism and Covert Attack.* Cambridge, MA: MIT Press, 2001.

Frist, M.D., Senator Bill. *When Every Moment Counts: What You Need to Know about Bioterrorism from the Senate's Only Doctor.* Lanham, MD: Rowman and Littlefield, 2002.

Garrett, Laurie. *The Coming Plague: Newly Emerging Diseases in a World Out of Balance.* New York: Farrar, Straus and Giroux, 1994.

Guillemin, Jeanne. *Anthrax: The Investigation of a Deadly Outbreak.* Berkeley, CA: University of California Press, 1999.

Harris, Robert and Jeremy Paxman. *A Higher Form of Killing: The Secret Story of Gas and Germ Warfare.* New York: Hill and Wang, 1982.

Harris, Sheldon H. *Factories of Death: Japanese Biological Warfare 1932-45 and the American Cover-Up.* London: Routledge, 1994.

Henderson, Donald A., Thomas V. Inglesby, and Tara O'Toole, Editors. *Bioterrorism: Guidelines for Medical and Public Health Management.* Chicago: AMA Press, 2002.

Hoffman, Bruce. *Inside Terrorism.* New York: Columbia University Press, 1998.

Kaplan, David E. and Andrew Marshall. *The Cult at the End of the World.* New York: Crown Publishers, 1996.

Knobler, Stacey L., Adel A. F. Mahmoud, and Leslie A. Pray, Editors. *Biological Threats and Terrorism: Assessing the Science and Response Capabilities.* Washington, DC: National Academy Press, 2002.

Lederberg, Joshua, Editor. *Biological Weapons: Limiting the Threat.* Cambridge, MA: MIT Press, 1999.

Levy, Barry S., and Victor W. Sidel, Editors. *Terrorism and Public Health.* New York: Oxford University Press, 2003.

Mangold, Tom and Jeff Goldberg. *Plague Wars: A True History of Biological Warfare.* New York: St. Martin's Press, 1999.

McDermott, Jeanne. *The Killing Winds: The Menace of Biological Warfare.* New York: Arbor House, 1987.

Miller, Judith, Stephen Engelberg, and William Broad. *Germs: Biological Weapons and America's Secret War.* New York: Simon and Schuster, 2001.

Osterholm, Michael T. and John Schwartz. *Living Terrors: What America Needs to Know to Survive the Coming Bioterrorist Catastrophe.* New York: Delacorte Press, 2000.

Piller, Charles and Keith R. Yamamoto. *Gene Wars: Military Control Over the New Genetic Technologies.* New York: Beach Tree Books, 1988.

Preston, Richard. *The Demon in the Freezer: A True Story.* New York: Random House, 2002.

Regis, Ed. *The Biology of Doom: The History of America's Secret Germ Warfare Project.* New York: Henry Holt, 1999.

Roberts, Brad, Editor. *Hype or Reality? The 'New Terrorism' and Mass Casualty Attacks.* Alexandria, VA: CBACI, 2000.

Spiers, Edward M. *Chemical and Biological Weapons: A Study of Proliferation.* New York: St. Martin's Press, 1994.

Stern, Jessica. *The Ultimate Terrorists.* Cambridge, MA: Harvard University Press, 1999.

ter Haar, Barend. *The Future of Biological Weapons.* New York: Praeger, 1991.

Thompson, Marilyn W. *The Killer Strain: Anthrax and a Government Exposed.* New York: HarperCollins, 2003.

Trevan, Tim. *Saddam's Secrets: The Hunt for Iraq's Hidden Weapons.* London: HarperCollins, 1999.

Tucker, Jonathan B., Editor. *Toxic Terror: Assessing Terrorist Use of Chemical and Biological Weapons*. Cambridge, MA: MIT Press, 2000.

Williams, Peter and David Wallace. *Unit 731: Japan's Secret Biological Warfare in World War II*. New York: The Free Press, 1989.

Zilinskas, Raymond A., Editor. *Biological Warfare: Modern Offense and Defense*. Boulder, CO: Lynne Rienner Publishers, 2000.

ADDITIONAL REPORTS

Carus, W. Seth. "Bioterrorism and Biocrimes: The Illicit Use of Biological Agents in the 20th Century," Working Paper, Center for Counterproliferation Research, National Defense University, Washington, DC, August 1998.

Centers for Disease Control and Prevention, "Update: Investigation of Bioterrorism-Related Anthrax and Interim Guidelines for Exposure Management and Antimicrobial Therapy," *Morbidity and Mortality Weekly Report*, Atlanta, GA, October 26, 2001.

Jernigan, Daniel B. et al. "Investigation of Bioterrorism-Related Anthrax, United States, 2001: Epidemiologic Findings," *Emerging Infectious Diseases*, October 2002.

Meselson, Matthew et al. "The Sverdlovsk Anthrax Outbreak of 1979," *Science*, November 18, 1994.

Parachini, John. "Anthrax Attacks, Biological Terrorism and Preventive Responses," CT-186, RAND, Washington, DC, November 6, 2001.

INTERNET RESOURCES

Centers for Disease Control and Prevention, "Anthrax":
http://www.bt.cdc.gov/agent/anthrax/index.asp

Federal Bureau of Investigation, "Amerithrax":
http://www.fbi.gov/anthrax/amerithraxlinks.htm

Lake, Ed, private investigator, "The Anthrax Cases":
http://www.anthraxinvestigation.com/

National Institutes of Health, "Anthrax":
http://www.nlm.nih.gov/medlineplus/anthrax.html

Occupational Safety and Health Administration, "Protecting the Worksite against Terrorism—Anthrax":
http://www.osha.gov/bioterrorism/anthrax/

Smith, Richard M., private investigator, "The Anthrax Investigation":
http://www.computerbytesman.com/anthrax/index.htm

University of California, Los Angeles, "Epidemiologic Information on Bioterrorism":
http://www.ph.ucla.edu/epi/bioter/bioterrorism.html

ACKNOWLEDGMENTS

This book was built on the courage, sadness, frustrations, hopes, and determination of many people. They include, foremost, the victims of the anthrax letters, the people who cared for them, and others who are dedicated to protecting this nation from bioterrorism in the future.

Among the anthrax victims and their families and friends, who generously shared their time and recollections with me, I thank Ernie Blanco, Dena Briscoe, Dave Cruz, Shirley Davis, Jim Harper, Teresa Heller, David Hose, David Hose, Jr., Johanna Huden, Qieth McQue, Maria Orth, Jyotsna Patel, Ramesh Patel, Allison Paliscak, William Paliscak, Leroy Richmond, Susan Richmond, Robin Shaw, Anna Rodriguez, Esperanza Vassello, and Norma Wallace. Friends and colleagues of Bob Stevens who spoke with me about him or about anthrax at American Media, Inc., include Bobby Bender, Joan Berkley, Carla Chadick, Martha Crompton, Ron Haines, Lee Harrison, Gloria Irish, Mike Irish, Martha Moffett, Sheila O'Donovan, Daniel Rotstein, Ed Sigall, Roz Suss, and Martha Warwick.

I thank three people in particular for the immense help they provided to me in the course of this project: Dr. Larry Bush, who persisted in his diagnosis of Bob Stevens's anthrax despite skepticism from colleagues; Dr. Nancy Connell, whose guidance on scientific issues was frequent, unstinting, and indispensable; and Dr.

D. A. Henderson, whose contributions to public health deserve recounting in a volume of their own.

Several scientists patiently explained to me various research and analytical techniques: Amol Amin, Sara Beatrice, Philip Cole, Nancy Connell, Mitchell Gayer, Paul Keim, John Kornblum, Philip Lee, Jessica Mann, Donald Mayo, and David Perlin. For explanations about the operation of the U.S. Postal Service and its connection to the anthrax letters I thank David Bowers, Dena Briscoe, Jim Harper, Thomas Day, William Lewis, Genevieve Mulvihill, Joseph Sautello, and Diane Todd. Several current and former agents of the Federal Bureau of Investigation were very helpful, including William Atkinson, Robert Blitzer, John Iannarelli, Randolph Murch, Judy Orihuela, and Christopher Whitcomb.

Terrorism analysts who shared their insights with me include Jeffrey Bale, Boaz Ganor, Bruce Hoffman, Michael Hopmeier, John Parachini, Jerrold Post, Peter Probst, and Jessica Stern. Similarly, experts in biological defense or biological arms control included Ronald Atlas, Kenneth Bernard, Charles Dasey, Edward Eitzen, Gerald Epstein, Elisa Harris, Marie Chevrier, Barbara Rosenberg, Richard Spertzel, Jonathan Tucker, and Alan Zelicoff.

Many public health officials offered the benefit of their experiences to me on matters related to the anthrax letters and biopreparedness. Among them, current or former officials from the Centers for Disease Control and Prevention: Philip Brachman, Julie Gerberding, James Hughes, Richard Kellogg, Jeffrey Koplan, Scott Lillibridge, Donald Millar, Stephen A. Morse, Stephen Ostroff, Tanja Popovic, Bradley Perkins, Jennita Reefhuis, Thomas Skinner, and Sherif Zaki; and from the National Institutes of Health, Anthony Fauci and Mary Wright.

State and local health officials included Joel Ackelsberg, Rex Archer, Eddy Bresnitz, George DiFerdinando, James Hadler, Timothy Holtz, Clifton Lacy, Marcelle Layton, Brian Letourneau, Jean Malecki, John Marr, Carol Moehrle, Lloyd Novick, Michael Osterholm, Ron Osterholm, Faruk Presswalla, James Spitzer, Christina Tan, and Steven Wiersma.

Other scientists or healthcare professionals who were helpful with issues related to biological defense and preparedness were Ken Alibek, Thomas Argyros, Sue Bailey, Richard Ebright, Michael Franzblau, Martin Furmanski, Margaret Hamburg, Martin Hugh-Jones, Thomas Inglesby, Roberta Lavin, Joshua Lederberg, Brendan McCluskey, Matthew Meselson, Stephen S. Morse, Tara O'Toole, Zarnaz Rauf, Monica Schoch-Spana, Shmuel Shapira, and Boaz

Tadmor. Physicians who treated or oversaw treatment of anthrax cases (confirmed or suspected) included Lydia Barakat, Larry Bush, Elyse Carty, Tyler Cymet, John Eisold, Mayer Grosser, Gary Kerkvliet, Norman Lee, Susan Matcha, Parvaiz Malik, Bushra Mina, Cecele Murphy, Michael Nguyen, Carlos Omenaca, Baksh Patel, Howard Quentzel, Jonathan Rosenthal, Michael Tapper, and Martin Topiel.

On matters concerning anthrax hoaxes, Harvey Berk, Chris Bosch, Sidney Clearfield, Robert Duffy, Carmen Fontana, Richard Heideman, Blake Kaplan, Reggie Lattimer, Dan Mariaschin, Mary McColl, Mark Olshan, Greg Ono, Luis Ortiz, Jason Pate, Reva Price, Rebecca Poedy, Dennis Roddy, Angie Stefaniak, Christine Vendel, and Clayton Lee Waagner were helpful.

For their valuable comments on the manuscript, I thank Emil Gotschlich, Paul Talalay, Scott Shane, and Jonathan Tucker. I thank Rebecca Allen for her cheerful acceptance of last-minute requests to transcribe hours of tape. For other varied, but highly valued, contributions, I am grateful to Kim Chiorella, Pat Clawson, Russell Feingold, Lawrence Feldman, Ralph Gomory, Theodore Greenwood, Pat Hallengren, Ron Krumer, Shalom Levin, Elizabeth McGuire, Mary Murphy, Wendy Orent, Laura Petrou, William Powanda, Hannah Rosenthal, Mary Ann Schierholt, Ann Silverman, and Susan Whyte Simon.

I am deeply indebted to John Thornton, my agent at the Spieler Agency, for his unfailing kindness and wisdom; to Jeffrey Robbins, my editor at the Joseph Henry Press, for his many valuable suggestions toward enhancing this book; and to the many others at the press for their support including Stephen Mautner, Heather Schofield, Ann Merchant, Jennifer Risko, Robin Pinnel, Jessica Henig, and Alvaro Rojas. For his editorial advice I am especially grateful to my dear friend, Richard Karlen, and for a lifetime of partnership and support, to my wife, Ruth Cole.

INDEX

A

B

C

D

E

M

N

S

T

U

V

W

Z